Alejandra Medina

Climate Financing in Ecuador

Alejandra Medina

Climate Financing in Ecuador

under the United Nations Framework Convention

ScienciaScripts

Imprint
Any brand names and product names mentioned in this book are subject to trademark, brand or patent protection and are trademarks or registered trademarks of their respective holders. The use of brand names, product names, common names, trade names, product descriptions etc. even without a particular marking in this work is in no way to be construed to mean that such names may be regarded as unrestricted in respect of trademark and brand protection legislation and could thus be used by anyone.

Cover image: www.ingimage.com

This book is a translation from the original published under ISBN 978-620-2-14273-1.

Publisher:
Sciencia Scripts
is a trademark of
Dodo Books Indian Ocean Ltd. and OmniScriptum S.R.L publishing group

120 High Road, East Finchley, London, N2 9ED, United Kingdom
Str. Armeneasca 28/1, office 1, Chisinau MD-2012, Republic of Moldova, Europe
Managing Directors: Ieva Konstantinova, Victoria Ursu
info@omniscriptum.com

Printed at: see last page
ISBN: 978-620-8-40059-0

CLIMATE FINANCE UNDER THE UNITED NATIONS FRAMEWORK CONVENTION IN ECUADOR

Summary

Climate change as a global problem requires an urgent global response. It does not respect borders and highlights social, economic, cultural and political problems. It also widens inequality gaps, conflicts and instability between countries. Climate finance is the cornerstone for the materialization of agreements, commitments and the fulfillment of climate goals. This research analyzes climate finance for REDD+ Ecuador under the UNFCCC, directed to PROAmazon. To evaluate the impact of climate finance, a climate investment analysis was conducted to determine the effectiveness, efficiency and equity (3E+) of the activities carried out with these funds, through the variables: deforestation rate, *stakeholder* participation and land tenure. After the conclusion of PROAmazonian activities in the period 2017-2023, the results show an increase in the deforestation rate, which is considered one of the highest in Latin America, caused by extractivist activities. The participation of indigenous peoples and communities as the managers of forest conservation and protection has been underestimated. Land tenure security has been a slow process and is threatened by the expansion of the extractive industry. The research concludes that the result of the application of climate finance has not achieved the expected results, and has been ineffective, inefficient and inequitable.

Key words: climate change, deforestation, effectiveness, efficiency, equity

Table of Contents

INTRODUCTION..4

CONCEPTUAL FRAMEWORK 11

METHODOLOGY 25

CLIMATE CHANGE, DEFORESTATION, REDD+ AND CLIMATE FINANCE 28

RESULTS..61

SHORTCOMINGS OF REDD+ IN TERMS OF EFFECTIVENESS, EFFICIENCY AND EQUITY 61

CONCLUSIONS...85

BIBLIOGRAPHY 88

Annexes 110

INTRODUCTION

The devastating effects of climate change are evident in the increase in the average temperature of the atmosphere and the oceans, alterations in the global water cycle, the increasingly frequent occurrence of extreme weather events, the reduction in the volume of snow and ice on glaciers and mountain peaks, and the rise in sea level, among others. The United Nations Framework Convention on Climate Change (UNFCCC) establishes a difference between climate change attributable to human activities that alter atmospheric composition and natural causes that cause climate variability (IPCC 2014), the two cases cause "variability", one of natural origin and the other anthropogenic. It is highly likely that human activity is the dominant cause of global warming.

Concentrations of greenhouse gases (GHGs)[3] are increasing rapidly reaching record concentrations. For example, carbon dioxide (CO_2) has exceeded 400 parts per million (ppm) compared to 280 ppm in the pre-industrial era (Berruezo and Díaz 2017). Social, economic and technological factors are recognized in the increase of GHGs, such as population growth, increased per capita demand for energy and resources, the use of inadequate technologies, among others. It is estimated that by 2030, 50% more food, 45% more energy and 30%

[3] Greenhouse gases (GHGs) are gases that trap heat in the atmosphere and their effect on climate change depends on the amount of gases, how long they remain in the atmosphere and how they affect global temperature (Romano, et al. 2018). CO_2 is the main GHG emitted by anthropogenic activities and constitutes around 65 % of global GHG emissions (IPCC 2014) in turn, three quarters of anthropogenic GHG emissions come from industrialized countries (IEA 2016)

more water will be needed (Molina, Carabias and Sarukhán 2017).

The planet has a great diversity of terrestrial and marine ecosystems that are the result of thousands of years of interaction and evolution of climatic and biotic factors, and play an important role in climate regulation. Ecosystems can be affected by natural causes such as hurricanes, volcanic eruptions and especially by anthropogenic causes such as deforestation, fires, intensive agriculture, industry, infrastructure, transportation, extractive activities, among others (Berruezo and Díaz 2017).

Approximately 300 million people live in the world's forests, 1.6 billion depend directly on these ecosystems, this estimate suggests that about one third have a close relationship and dependence on forests and forest products and 7.8 billion people need them to live, i.e. all of humanity. Between 17% and 20% of global annual CO_2 emissions come from deforestation and forest degradation (FAO 2020).

Forest ecosystems participate in the fight against climate change by absorbing carbon through the process of photosynthesis, which is stored in the trunk, branches, roots and soil. The boreal forests of the far north and tropical rainforests are home to enormous biodiversity, with more than 60,000 different tree species providing habitats for 80% of amphibian species, 75% of bird species, 68% of mammals and 60% of vascular plants (FAO 2020). The conservation of a large percentage of biodiversity depends on the care and good use of forests. With this premise, a global approach to forest protection and conservation is needed between the governments of the different nations of the world, international organizations and civil society.

Land use change, deforestation and forest degradation are a global problem and are the main sources of carbon emissions. Changes in vegetation cover in forest ecosystems have caused temperature increases of 0.23°C in the period 2000-2015. An area without vegetation cover absorbs more radiation and increases in temperature; therefore, there is a close relationship between deforestation and global warming. To this increase is added the warming caused by CO_2, which produces a double impact: first the local one, on the deforested area and its repercussions on ecosystems, and second the global one, adding more heat to climate change (Duveiller, Hooker and Cescatti 2018).

In higher elevations such as boreal regions, the loss of trees rapidly causes the surface albedo to rise. Whereas in tropical areas with permanent and broadleaf tree species, deforestation by agriculture and/or livestock increases surface albedo, but eliminates the thermal regulation produced by forests. Replacing permanent and broadleaf rainforests with another type of vegetation triggers an increase in temperature; changing to another type of vegetation cover is not compensated by planting trees in an equivalent area at other latitudes. Therefore, if any deforestation process is negative, tropical rainforests require more attention and effort to conserve them (Duveiller, Hooker and Cescatti 2018).

For this reason, the UNFCCC created the Reducing Emissions from Deforestation and Forest Degradation mechanism, which includes conservation, sustainable management and enhancement of forest carbon stocks (REDD+). The mechanism seeks to slow, halt, reverse forest loss, ensure biodiversity conservation and the well-being of communities living in forests (UNEP 2018b).

The need for a solution to deforestation and forest degradation has prompted an international discussion on climate finance, even though to date no consensus definition and agreement has been reached on the subject. Financial resources are not only necessary to contribute to the sustainable development of developing countries, but also to adapt to adverse effects and reduce climate impacts. In this context, financing for Latin America in the period 2003-2020 has been concentrated mainly in two countries: Brazil (USD 1.179 billion) and Mexico (USD 555 million), which together have received 35% of the total of USD 5 billion in climate financing for the region. Mitigation activities such as forest protection and reforestation have received the most funds compared to adaptation activities, with USD 3.4 billion and USD 670 million respectively. In 2020, the Green Climate Fund (GCF) was the largest provider of climate finance for the region with USD 1,906 million (Watson, Schalatek and Evéquoz, 2022) and allocated USD 458 million for Payment for Results (PPR)[4] of REDD+ (Watson and Schalatek 2021). PPR are financial funds that REDD+ earmarks for the results of climate activities and are subject to GHG measurement, reporting and verification at the national level (UNEP 2018b).
Ecuador is a low per capita income country, marked by high income

[4] Payment for Results are results-based REDD+ actions and are subject to GHG measurement, reporting and verification at the national level (UNEP 2018b).

inequality[5] estimated at 49.3% (World Bank 2021a), euphemistically called developing, vulnerable to the adverse effects of climate change. It has a regulatory framework that considers mitigation and adaptation responses at the national level through the National Climate Change Strategy 2012-2015 (ENCC), which is a public policy instrument that guides actions at the national and sectoral levels, the Organic Environmental Code (COA) and its Regulations (RCOA) that address climate change and its financing.

Ecuador is committed to promoting sustainable production systems and encouraging the restoration of deforested and degraded areas in order to conserve forests through access to international financing and with the participation of various actors from the public and private sectors, indigenous nationalities and non-governmental organizations (NGOs). Currently, the main entity managing climate change financing is the Ministry of Environment, Water and Ecological Transition (MAATE)[6] as the recipient of financial resources and executor of most projects.

The country has proposed to prioritize the reduction of the deforestation rate through a series of national policies. In 2008, MAATE launched the Socio Bosque Program (PSB) as a program for the conservation of native forests and moorlands that promotes the reduction of GHGs in

[5] The Lorenz curve and the Gini coefficient are tools used in the field of economics to measure income inequality in a population or society (González 2020).

[6] In March 2020, through Executive Decree No. 1007, former President Lenín Moreno ordered the merger of the Ministry of Environment (MAE) and the Secretariat of Water (Senagua) creating the Ministry of Environment and Water (MAAE) and in June 2021, through Executive Decree No. 59, former President Guillermo Lasso renamed this State portfolio as the Ministry of Environment, Water and Ecological Transition (MAATE).

the country's forests.

in exchange for a financial incentive. The PSB favored Ecuador for being one of the Latin American countries to receive PPR through REDD+ (Nepstad, et al. 2019).

In this framework, evaluating climate finance through REDD+ requires analyzing the economic resources received up to 2023, which amount to USD 145.8 million and which have been earmarked for reducing deforestation and forest degradation. These resources come from GCF funds, Global Environment Facility (GEF,) and from the co-financing[7] reported between MAATE, Ministry of Agriculture and Livestock (MAG) [8][9] , United Nations Program (UNDP) and Food and Agriculture Organization of the United Nations (FAO).

The purpose of the research is not an evaluation of changes in carbon emissions, but an analysis of the effectiveness, efficiency and equity (3E+) of REDD+ financial funds directed to the Amazonian Integral Program for Forest Conservation and Sustainable Production (PROAmazon) through the variables: deforestation rate, *stakeholder*

---

[7] "Co-financing is a practice in which multiple entities finance the same project. Co-financing may be provided by the project developer or by external entities. A robust cofinancing scheme (whether in-kind or cash) is evidence of broad interest in the project from a diversity of relevant stakeholders and is therefore an important feature of project design" (ICLEI 2020). It is also defined as a "loan granted to developing countries by commercial banks and other lending institutions, in association with the World Bank and other multilateral development banks" (ECLAC 1989).

[8] In March 2017, through Executive Decree No. 207, former President Lenín Moreno changed the name of the Ministry of Agriculture, Livestock, Aquaculture and Fisheries (MAGAP) to Ministry of Agriculture and Livestock (MAG).

[9] The term is used in English because there is no satisfactory translation in Spanish.

participation[1] and land tenure.

CONCEPTUAL FRAMEWORK

GHG emissions from deforestation and forest degradation represent approximately 20% of the total annual global emissions of $CO_{(2)}$ (FAO 2020). For this reason, in recent years, forests have been the focus of attention in international climate debates. As a collaborative initiative of the United Nations (UN), the Reducing Emissions from Deforestation and Forest Degradation (REDD) mechanism was created to encourage developing countries to protect, manage and make better use of forest resources in the fight against climate change. The main idea was to conserve standing forests for their carbon sequestration capacity, assigning them a financial value and thus avoiding logging. This action made it possible to quantify carbon, the final stage of REDD, which included the payment of compensation for standing forests by developed countries to developing countries (Rudel, et al. 2005). In other words, carbon reductions are quantified and compensation payments are made by developed countries to developing countries.

Under this scheme, REDD discussions focused on the creation of a mechanism for payments for environmental services (PES)[10] to developing countries, at the national and international level. At the international level, service buyers make a payment (voluntary market) to service providers (governments, national entities) for the reduction of GHG emissions as a consequence of reducing and avoiding

[10] Payments for environmental services have certain advantages such as creating incentives for forest owners and users to manage forests better and cut down fewer trees. PES mechanisms compensate carbon sink (forest) rights holders committed to forest conservation as a lucrative alternative (Angelsen, Kanninen, et al. 2010).

deforestation and forest degradation. At the national level, governments or other actors (service buyers) pay subnational governments or local landowners for GHG reductions (Angelsen and Wertz- Kanounnikoff 2009).

Although various forms of PES have been implemented, there have been difficulties in their application, so it has been necessary to create institutional and forest governance structures to manage payments and information with the objective of linking local PES systems with global and national REDD systems. One of the challenges of REDD has been to ensure that the payments made through the various systems are effective, efficient and equitable (Angelsen and Wertz-Kanounnikoff 2009).

For political and technical reasons, the concept of REDD was expanded to REDD+. The REDD+ mechanism was born as an innovative, cheap, easy and quick implementation tool, with a broader vision, not only to reduce GHG emissions but also to generate a greater financial flow to reduce poverty and conserve biodiversity. This postulate has been the subject of much controversy, initially revolving around the global architecture of REDD+ and how to include it in a post-2012 climate agreement, later the debate focused on national and local actions.

1. Criteria 3E+

According to the Stern Report (2006) on the Economics of Climate Change, reducing GHG emissions by reducing deforestation would cost on average between one and two dollars per ton of CO_2, an apparently cheap estimate compared to other mitigation options. The Stern report introduced for the first time the concepts of effectiveness, efficiency and equity known as the 3E criteria. This approach broadened the perspective towards a mechanism that generates sufficient incentives to curb deforestation and consequently reduce global GHG emissions.

The objective of the report was to design viable policies and projects to achieve climate effectiveness, cost efficiency and results in terms of equity (3E), by adding the symbol "+" to the three criteria, reference was made to co-benefits in biodiversity: poverty reduction, generation of sustainable livelihoods, governance, community rights and participation, tenure and enhancement of non-carbon ecosystem services. These criteria, known as the 3E+ (Angelsen, Brockhaus, et al. 2013), help to evaluate GHG reduction schemes at the lowest possible cost and contribute to sustainable development, since ideally a REDD+ project should comply with the 3E+ (Angelsen and Agrawal 2009).

In this context, an effective design and implementation of REDD+ requires a set of policies that include institutional reforms in forest governance, tenure, decentralization and community forest management. Despite the efforts made, it has not yet been possible to stop deforestation from advancing because the main drivers of deforestation and land trend problems are not addressed by considering the forestry sector in isolation. Policy formulation should aim to reduce agricultural rent in forest areas, increase the value of standing forests and help forest users capture that value, directly regulate land use and land tenure; these policies should work together to achieve REDD+ outcomes in terms of the 3E+ (Sunderlin, Larson and Cronkleton 2010).

1.1. Effectiveness

Effectiveness refers to the amount of GHG emissions reduced as a result of REDD+ mechanism activities. It depends on several factors such as political feasibility, governance and commitment of countries to implement the mechanism. It also depends on other considerations such as leakage control or avoidance, corruption, permanence, accountability and the extent of the main drivers of deforestation and degradation (Angelsen, Kanninen, et al. 2010).

It focuses on the capacity to achieve the planned objectives, regardless of the resources allocated and helps the analysis of the sustainability of the mechanism's actions and effects, including those in neighboring areas (Nepstad, et al. 2019). REDD+ as a PPR mechanism is considered effective when there is economic aid despite skepticism, contradictions and aid conditionality (Paul 2015, Angelsen 2017a). This requires countries to report their results on GHG emission reductions through the Monitoring, Reporting and Verification System (MRVS)[11] , quantifiable and verifiable information to avoid data leakage. The success of REDD+ depends on a solid institutional structure through laws, policies and strategies on environment, climate change and sustainable development (Kambire et al. 2016).

REDD+ environmental negotiations have focused on reducing deforestation in tropical rainforests, this could be more effective.

---

[11] The Monitoring, Reporting and Verification System allows once the baseline with Reference Levels is established to monitor GHG emissions and establish reduced carbon accounting through REDD+ implementation (UNEP 2018b).

if there were community-based forest management[12] "to address emissions resulting from forest degradation than those from deforestation, and that the effectiveness of community-based forest management could be maximized in dry tropical forests" (Matta and Meins 2012).

Efficiency

Efficiency focuses on GHG emission reductions at the lowest possible cost and time . Several costs are considered within REDD+: capacity building costs (scheme design, technical infrastructure, training); operating costs (monitoring, forest policy enforcement and land tenure); implementation costs borne by landowners, managers and forest users. "All of these, except for compensation and rent, are transaction costs" (Angelsen, Kanninen, et al. 2010).

On the other hand, environmental governance and the performance of national policies on deforestation and forest degradation are relevant. In order to maximize their economic potential, countries develop policies to promote strategic sectors such as agriculture, mining, energy and especially exportable goods, and apply their own guidelines that promote socioeconomic development without considering environmental protection or the negative effects of deforestation and forest degradation. Meanwhile, country policies that promote sustainable development and environmental protection are not efficient

---

[12] "Sustainable forest management involves, among other things, keeping records of flora and fauna, monitoring ecologically important forest areas, conducting reduced impact logging, forging public-private partnerships, and equitably distributing benefits among stakeholders" (Matta and Meins 2012).

without an institutional framework and the good will of all sectors and actors. The framework and the governance system are two parameters that determine advantages or limitations in the application of the 3E+ criteria (Kambire et al. 2016).

Given this context, it is timely to consider the political and socioeconomic dimensions of REDD+ performance [13] ranging from the global to the local level such as PPR, SMRV, co-benefits and community participation. While REDD+ performance includes the outcome, it is a dimension that transcends results and payments (Ramos et al. 2007). The lack of performance of the mechanism may be due to flaws in the very essence of the instrument that did not consider the environment in which REDD+ was to operate, where powerful actors have hindered the initial vision in order to maintain the status quo. Similarly, the evolution from PES to conservation efforts have been inefficient because of flaws in their conceptual design and market-based instruments (Angelsen, Duchelle, et al. 2017). To the extent that climate change becomes more pressing, carbon markets function better and problems are solved in reference to the effectiveness and efficiency of REDD+ as a forest governance mechanism and policy process, it could work (Angelsen, Martius, et al. 2019); the PPR approach does not guarantee that REDD+ will be an effective, efficient or equitable mechanism, it requires linking PPRs to the contexts in which such outcomes are defined and agreed under conditions of social and

[13] REDD+ performance is the act or process of performing a function and is evaluated within a public policy, project or program when the established objectives are achieved (Ramos et al.2007).

political acceptance (Wong, Luttrell, et al. 2019).

1.3 Equity

Equity within REDD+ proposals includes objectives that are not linked to climate change but have a connotation with social and environmental dimensions such as benefit sharing, livelihoods, poverty reduction, land tenure, protection of the rights and participation of indigenous peoples and communities, incorporation of a fairer gender perspective, biodiversity, among others (Angelsen, Kanninen, et al. 2010).

This criterion can be approached from different dimensions: from a global perspective, it is related to the fair distribution among countries based on the level of poverty and the capacity of poor countries to participate in REDD+ processes; from a national perspective, from a fair distribution within the same countries, for example, the distribution of costs and benefits in local governments and the national government; and from a subnational perspective, from the perspective of indigenous peoples and communities, in the recognition of their traditional practices and rights as well as in the processes of inclusion in REDD+ decision making. Therefore, the participation of indigenous peoples and communities, diverse social groups and civil society are considered important for the preparation and implementation phase, legitimacy and design of the REDD+ mechanism (Chhatre et al. 2012).

When national processes address local equity issues, it is assumed that they are processes delineated and led by each country according to its reality; however, the presence of International Organizations (IOs) such as the World Bank (WB), the UN and others, can result in extensive administrative processes that demand a large amount of technical resources (Romijn et al. 2015). Similarly, instruments to address equity such as social and environmental safeguards (ESS) are generally reduced to administrative, monitoring and reporting exercises. This can result in negotiations becoming detached from national objectives in reference to land, indigenous peoples and communities, forests and end up in weak integration with important sectors of change (Dawson et al. 2018). Given the above, the equity criterion may have difficulties in the distribution of costs and benefits, in decision-making procedures and recognition of identities and values that are determinant for the fulfillment of ecological objectives (Myers et al. 2018).

When analyzing the 3E+ criteria in a project, conflicting results can be observed. For example, a REDD+ project may yield good results at a large scale at relatively low cost but generate an increase in inequality in land ownership. As another example, a community-driven project to strengthen local land tenure rights may achieve equity gains but be costly and long-lasting (inefficient) (Springate-Baginski and Wollenberg 2010).

2. Variables of analysis

Each country's forest context is unique, the causes of deforestation and degradation are different, and development processes are also different. Therefore, considering the diversity of national circumstances, REDD+ strategies must respond to the needs of each country.

To evaluate the 3E+ criteria for climate finance, the deforestation rate was chosen, which in itself is important because developing countries with high rates of deforestation and forest degradation increase the environment for REDD+. In addition, the rate of deforestation allows comparing the change in forest area with and without REDD+, which makes it possible to analyze the effectiveness and efficiency of the mechanism in reducing GHGs (Angelsen, Kanninen, et al. 2010).

Stakeholder participation and land tenure were selected because compensation and safeguards related to indigenous peoples and communities have been postponed in climate investment, putting the interest of the mechanism before market-based systems. These variables highlight their importance within forest management to quantify the equity criterion (Angelsen, Kanninen, et al. 2010).

2.1. Deforestation rate

Deforestation rate refers to the permanent change in forest area between one period of time and a subsequent period of time caused by humans (Takaki 2010). REDD+ addressed economic incentives to change the approach of forest holders. Protecting forests means forgoing income by avoiding exploitation of this resource; in other words, forest conservation is more profitable than logging, agriculture and cattle ranching in order to receive PPR. Under this scheme, landowners conserve forests because they will have higher income, this connotation marked the difference from previous actions for forest conservation (Sunderlin and Atmadja 2009).

"Deforestation happens because it is profitable for someone. There's a lot of money in cutting down trees, mostly to turn land into agricultural fields. And the idea that REDD+ should change the equation by making a living tree more valuable than a dead tree will cost a lot of money if it is really to be done" (Angelsen 2020).

Lessons learned have shown that economic resources alone do not stop deforestation, REDD+ has not addressed the real causes of large-scale deforestation and is not contributing to climate protection because stopping deforestation is not a quick, easy or cheap process. Moreover, behind the postulate of the mechanism, one detects the purpose of camouflaging the intentions of developed countries in protecting tropical forests by transforming subsidies intended for "development aid" into loans to climate projects and programs (Kill 2017)

2.2. Stakeholder participation

Stakeholder participation refers to the actors involved in the REDD+ Working Group (MoW), which includes the National REDD+ Authority, indigenous peoples and communities, Afro-Ecuadorians, Montubios, academia, civil society and the private sector.

The concept of community participation can be approached from various perspectives: in the political sphere as a way to achieve power, social development or for democratic purposes, in the economic sphere to obtain some material benefits, and in the social sphere it is related to processes in which people mobilize to achieve objectives with a view to meeting certain needs and producing social changes. Community participation concentrates these definitions and is summarized as an organized, inclusive and autonomous process, oriented towards collective and individual transformations, where people with varying degrees of commitment share values and objectives (Montero 2004).

Additionally, it encompasses other aspects such as its inclusive character due to its interest in achieving a goal, as well as integrating several activities to reach a common objective (Sánchez 2000), its political character by forming citizenship and strengthening civil society (Montero 2010; Clary and Snyder 2002).

Under this premise, multisectoral dialogue is essential among the actors involved such as: local governments, decentralized autonomous governments, citizens in general, international development agencies, private banks, NGOs, academia, technical support agencies, indigenous peoples and communities, among others, with a view to jointly promoting and implementing support initiatives, common goals and specific objectives to generate transformational changes (Ulloa 2013).

Indigenous peoples and communities are key actors for the achievement of objectives such as reducing emissions from deforestation and forest degradation and for the use, sustainable management and conservation of forest resources (Angelsen and Agrawal 2009). Because of their role, communities should have preferential or differentiated treatment because they have always lived in the territory and their presence should be reflected before any political discussion as direct actors in forest conservation (Ulloa 2013).

Indigenous peoples and communities as active agents stand out as holders of knowledge, ancestral and traditional knowledge, actors willing to adapt to the changing environment and protectors of forests due to their close relationship with nature; their role goes beyond this, as actors who build solid evidence of their protagonism. Protecting the rights of communities is the greatest guarantee in the face of forest resilience; it is to support their position and legitimacy as protectors of the biodiversity of their territory, the conservation of threatened ecosystems and the restoration of their degraded lands. To enforce their rights they have established alliances highlighting their presence that supports their position and legitimacy (Lozano 2018).

At the transnational level, the political mobilization of indigenous peoples and communities has occurred with the incorporation of territories into mechanisms for the conservation of forest resources and ecosystem services, which has modified their relationships between the transnational and the local focused on climate change (Ulloa 2013).

REDD+ policies generally allude to communities as potential beneficiaries and agents for implementation, community-based approaches have been part of environmental governance initiatives. They are recognized as actors responsible for managing local REDD+ initiatives (Skutsch and Turnhout 2018). The wording of the mechanism states that it relies on communities through participatory processes and "free, prior and informed consent" for its intervention. However, in practice, decision-making processes are rarely "free", hardly "prior" and little "informed" and hardly seek some form of democratic "consent" or even "consultation" (Ece, Murombedzi and Ribot 2017).

Forest-dependent communities are not involved in REDD+ projects, and the potential impacts of the mechanism disrupt their livelihoods, socio-cultural systems, food security, encourage the introduction of monocultures, the presence of powerful actors and the illegal acquisition of land in different ways. REDD+ is also described as a mechanism that renegotiates the relationship of peoples with their natural space by monetizing nature (Macchi, et al. 2008). Currently, for communities living in the forest is a challenge, their rights have been violated and they have become more sensitive to climate change (Bayrak and Marafa 2016).

2.3. Land tenure

Tenure refers to "the relationship, legally or customarily defined, between people, as individuals or groups, with respect to land" (FAO 2003). It is also understood as the set of relationships, systems and rules that govern the rights to use land and forest resources; it determines who can exploit, sell, exclude from the use of the land and the natural resources found on it; it also establishes the responsibilities and limitations related to the rights of use and exploitation (FAO 2016).

The lack of certainty in tenure rights leads indigenous peoples and communities to make more intensive and temporary use of the land through short-cycle planting and intensive livestock raising. When communities develop itinerant agriculture, they exhaust the soil and move to other areas because they do not have the right of belonging and security that comes with the right to land ownership, which affects communities that are already affected by other causes such as poverty, lack of employment and marginality (FAO 2016).

Tenure is not necessarily synonymous with ownership or dominion. In some cases tenure coincides with ownership but this is not a condition;

in others, indigenous peoples and communities do not have a recognized right to a land, do not have registered property titles but use certain territory for some generations, have their governance dynamics and internal organization; although, they do not have dominion from a legal point of view, they have the benefits of tenure (FAO 2016).

Land tenure is important in REDD+ planning and implementation; it forms the basis on which projects and benefit sharing are built. Tenure and rights are relevant to the development of REDD+ safeguards. The mechanism promotes forest investment and management in areas that are generally remote from urban centers and in locations that are difficult to access, lack of legal security over tenure is one of the main brakes for investment; therefore, clarifying and providing security over tenure rights is the first step in the REDD+ readiness process (FAO 2016).

In this framework, one of the conditions for implementing REDD+ globally has been to have security over the property rights of landholders to access financial resources. The mechanism has provided opportunities to clarify local tenure rights in intervention areas to secure their activities; they have allocated resources to incentivize the acceleration of forest tenure reforms in favor of their interests (Hubert 2014). For example, Brazil has a record of implementing land tenure reforms, with much of its forests under communal ownership or allocated for community use, including efforts to geo-reference small and medium-sized properties for inclusion in REDD+ projects (Dewan 2011).

METHODOLOGY

The starting point is climate finance under the UNFCCC through the GCF and GEF funds directed to PROAmazonia, in the commitment to repay Ecuador for conserving forests. The information of co-financing from other sources, is not the object of analysis of the research, but it is mentioned to complement the information of the total climate finance for the period 2017-2023.

The study is based on a variety of bibliographic sources that include scientific articles, climate finance reports, GCF and GEF project reports, semi-structured interviews with academics, environmental specialists and climate finance experts (Table 1) aimed at obtaining data and information on the actions carried out within REDD+ projects. The interviewees were selected based on their participation in the REDD+ Working Group (MoW), as well as their experience in environmental matters, and the technical and socioeconomic knowledge that supports the implementation of the projects in the territory. The response from the interviewees was a set of unstructured, varied and critical data, each with their own perspective; information that was given a structure and allowed for a broader analysis.

To deepen the analysis of the criteria effectiveness, efficiency and equity (3E+), a study of the following variables was carried out: 1. deforestation rate; 2. stakeholder participation; and 3. land tenure. An analysis of the variables was elaborated using relevant literature and information from interviews; a complex task considering that the dynamics of these variables are phenomena that have occurred under heterogeneous social, political, economic and institutional contexts.

Entrevista a Expertos

FECHA ENTREVISTA	ENTREVISTADO	CARGO	REPRESENTANTE	CÓDIGO
07/07/2021	Arild Angelsen	Profesor de Economía en la Norwegian University of Life Sciences (NMBU)	Academia	A1
14/06/2021	Carolina Rosero	Gerente de Políticas Ambientales de Conservación Internacional	ONG	O1
09/06/2021	Cristina García Soto	Oficial de Programa de Bosques y Agua de WWF	ONG	O2
15/06/2021	David Romo Vallejo	Director del Programa de Diversidad Etnica de Universidad San Francisco de Quito (USFQ)	Academia	A2
29/06/2021	David Yedra	Director del Gestión Ambiental de GAD Provincial de Pastaza	GAD	G1
17/08/2021	Duval Llaguno Ribadeneira	Especialista en Recursos Naturales del Banco Interamericano de Desarrollo	BID	B1
29/06/2021	Francisco Moscoso Silva	Especialista Técnico en Monitoreo y Seguimiento de PROAmazonia	PROAmazonía	P1
17/06/2021	Jaime Toro Guajala	Director de Naturaleza y Cultura Internacional	ONG	O3
05/05/2021	Jessica Gallegos	Especialista de Mitigación de Cambio Climático del MAATE	MAATE	M1
18/08/2021	Manuel Shiguango	Técnico Territorial CONFENIAE / ONU REDD+	CONFENIAE / ONU REDD+	C1
18/08/2021	Patricia Serrano	Gerente de PROAmazonía	PROAmazonía	P2

Table 1. Interviews with academics and experts Prepared by the authors.

For the deforestation assessment, the national deforestation rate was reviewed through a comparative historical analysis from 1990 to 2022 to examine the results achieved by GCF and GEF funded projects. The deforestation results of projects under these funds cause some uncertainty due to factors such as reduced data information and updated figures. This variable allows the effectiveness and efficiency of REDD+ activities to be quantified.

Regarding the participation of stakeholders, a mapping was carried out on the intervention of the actors involved (Annex 1), aimed at to obtain data and information on the actions carried out by PROAmazonia. The selection of the interviewees was based on their participation in the MoT, their experience in environmental matters, and the technical and socioeconomic knowledge that supports the implementation of the projects in the territory. The research analysis focuses on the participation of indigenous peoples and communities as a strategic partner for the implementation of REDD+ actions.

In order to evaluate land tenure, official and academic literature was reviewed. In addition, it was reinforced with the interviews conducted. This variable was considered because land rights are a prerequisite for accessing REDD+ benefits. Land tenure is the basis on which projects and benefit sharing are consolidated. These last two variables, *stakeholder* participation and land tenure, allow equity to be quantified.

CLIMATE CHANGE, DEFORESTATION, REDD+ AND CLIMATE FINANCE

The chapter discusses the problems that climate change has generated in the global context and the need for climate cooperation and financial participation by developed and developing countries to build inclusive, low-carbon and climate-resilient economies. The distribution of forests and their importance as ecosystems that generate multiple benefits and their direct relationship with climate change are analyzed. With this premise, it briefly addresses the climate negotiations that have taken place within the framework of the UNFCCC aimed at REDD+ as a financial mechanism that promotes voluntary efforts by developed countries to reduce GHGs and increase forest carbon stocks. It analyzes the climate finance resources committed under the UNFCCC over the last decade to reduce deforestation and forest degradation in developing countries.

1. Climate change

Climate change is currently one of the most relevant challenges facing humanity. According to the UNFCCC (2014a) it is a challenge that must be met through concrete actions to achieve the Sustainable Development Goals (SDGs) and compliance with the Nationally Determined Contributions (NDCs), which are the national initiatives of each country aimed at reducing GHGs in light of the Paris Agreement (PA). The challenge calls for joint action with climate investment to promote sustainable, low-carbon and climate-resilient environmental development, especially for the benefit of the most vulnerable countries (Hirsch 2018).

Climate change is defined as "change in climate attributed directly or indirectly to human activity that alters the composition of the global atmosphere and that is in addition to natural climate variability observed over comparable time periods" (UNFCCC 1992). Also, "it may be due to natural internal processes or changes in external forcing, or to persistent anthropogenic changes in the composition of the atmosphere or in land use" (IPCC 2013). In other words, it is a phenomenon caused by the accumulation of GHGs in the atmosphere resulting in an increase in average temperature that has led to the disruption of the climate system.

GHGs are gases that accumulate in the atmosphere, absorb infrared radiation from the sun, are necessary to maintain the temperature of the planet; however, human activity has increased their production altering the natural balance, causing the increase of global temperature, alteration of the radiant energy flow in the atmosphere (radiative forcing) and variation of the energy balance on the earth's surface (climate forcing) (Romano, et al. 2018). Three-quarters of CO_2 comes from industrialized countries (Xu, et al. 2017). In 2018, CO_2 emissions reached unprecedented levels, an emission of 37 billion tons of CO_2 was recorded from the burning of fossil fuels, this is 3.1 % more compared to the previous year (OECD 2019).

The challenge is to combat climate change. Among the global commitments to deploy financial resources to build inclusive low-carbon and climate-resilient economies, the available financing and the capacities to manage these resources differ from country to country. Developed countries have internal capacities to produce and use resources, while most developing countries require external financial resources to mitigate and adapt to climate change (UNDP 2012).

2. Forests

The distribution of forests on the planet is neither uniform nor equitable in reference to world population or geographic location. More than half of the world's forests are located in five countries: Brazil, Canada, China, the United States and Russia cover 31% of the world's total land area, representing 4.06 billion hectares (Mha) (Figure 1). Approximately half of the forest area (49 %) is almost intact and more than one third (34 %) represents primary forests, where there is no evidence of human activity and ecological processes have not undergone notable disturbances (FAO 2020). In addition, tropical areas have the largest percentage of the world's forests (45%) (FAO 2020a).

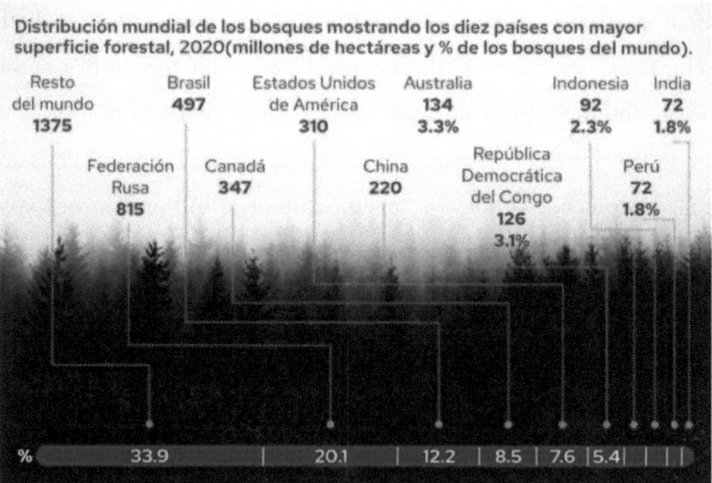

Figure 1. Forests of the world
Source: FAO 2021 Own elaboration

Carbon is in the environment in different forms such as biomass, litter, animals that return it to the soil through their waste or when they die; also, in the form of fossil fuels, rocks and in the atmosphere. The absolute amount[14] retained in the forms in which carbon is found are reserves that vary due to deforestation or degradation, these variations are called fluxes, so that carbon flows between the different reserves (FAO 2003a).

Based on this premise, forests and climate change have a direct relationship. Changes in climate affect forests and vice versa; increasing temperatures, changing rainfall, river and other extreme weather events affect forests. This complex interrelationship needs to be addressed in a comprehensive and innovative way, environmental degradation needs to be tackled and unsustainable economic growth needs to be curbed. "Forests are a nature-based solution to many sustainable development challenges" (FAO 2020).

2.1. Deforestation and global forest degradation

The planet's forest area is decreasing, although the rate of net loss has slowed. In the period 1990-2020, a loss of 178 Mha. of forest is accounted for as a consequence of reduced deforestation in several countries, increased afforestation in others and natural expansion of forests. The net loss rate for the period 1990-2000 is estimated at 7.8

[14] Not all forests store the same amount of carbon; in principle, carbon sequestration is determined by the amount of biomass; however, the amount sequestered depends not only on biomass but also on the type, age of the forests and other vegetation (Biomass + forest type and other variables = CO_2 stored). In general terms one ton of biomass is equivalent to half a ton of carbon (UNEP 2018).

Mha. per year, for the period 2000-2010 it decreased to 5.2 Mha. per year and for 2010-2020 it decreased to 4.7 Mha. per year (Figure 2) (FAO 2020a).

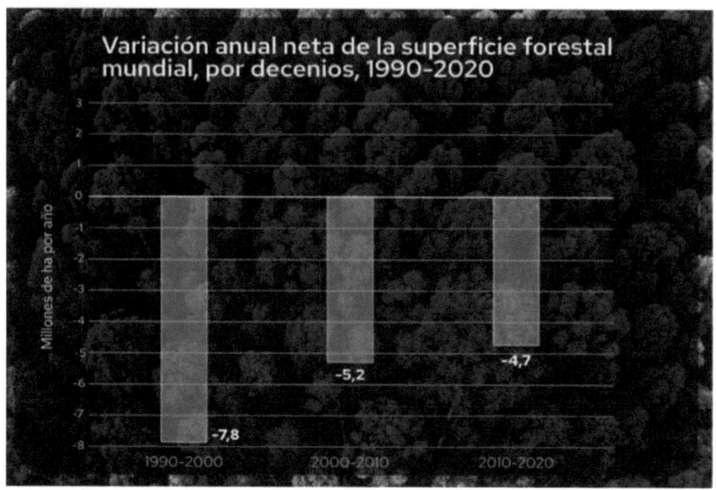

Figure 2. Net annual change in forest area
Source: FAO 2021
Own elaboration

Given this premise, it is necessary to establish the difference between deforestation and forest degradation. Deforestation is "the conversion of forests to another type of land use or the permanent reduction of canopy cover, below the minimum threshold of 10 percent" (FAO 2015) a process that implies loss or constant decrease of forest cover and transformation to another land use, for example: cropland, pasture, forestry, settlement, wetlands, etc. Forest degradation is "the decrease in the capacity of the forest to provide goods and services" (FAO 2015) in other words, it is not a decrease in forest area but in the quality of its condition, a process that implies the loss of carbon stocks.

It is estimated that 424 Mha. of forests on the planet are destined for the conservation of biodiversity; however, this area has decreased over the last ten years (FAO 2020a). More than 100 Mha. of forests are frequently threatened by forest fires, pests, droughts and adverse weather events. Large-scale agricultural and commercial expansion have been a constant cause of forest loss. Unfortunately, forests face many disturbances that significantly affect their vitality and reduce their capacity to provide ecosystem goods and services (FAO 2020).

The consequences are evident in the decrease of the total forest carbon stock. Most of it is found in living biomass, which represents 44%, and soil organic matter, which is equivalent to 45%; the remaining difference corresponds to dead wood and leaf litter. In 1990, the total carbon stock was estimated to be 668 gigatons (Gt), but by 2020 it had dropped to 662 Gt. For the same period, carbon density experienced a brief increase from 159 to 163 tons per ha (FAO 2020a).

1.1. UNFCCC and forestry mechanisms

Over the last century, climate change has received special attention on international, national and local agendas. The UNFCCC was based on one of the most successful multilateral environmental treaties, the Montreal Protocol of 1987, which called on member states to participate in human security despite the lack of current scientific evidence (Blobel et al. 2006).

In 2005, at the Conference of the Parties (COP) No. 11 in Montreal, the Coalition of Rainforest Nations (CNBT) proposed an offset mechanism called Reducing Emissions from Deforestation (RED) based on the concept of compensated reduction and a cost-effective mitigation option (Fry 2008) and its contribution to mitigation would provide important co-benefits for ecosystem services and biodiversity. This was the first approach of the Parties to the UNFCCC to one of the most critical issues of the climate negotiations and the beginning of the effective implementation of the Principle of Common but Differentiated Responsibilities (PRCD)[15] . Voluntary participation and positive incentives would enable developing countries to meaningfully collaborate in a climate regime without undermining their development capacity (Rudel, et al. 2005).

Initially, negotiations focused on RED from deforestation as a mechanism to create incentives to protect, optimize management and wise use of forest resources in developing countries, and later extended to forest degradation (REDD). In 2008, in response to the need for technical assistance and capacity building, multilateral institutions such as the WB's Forest Carbon Partnership Facility (FCPF) were created as a global strategic alliance to assist developing countries in their efforts to reduce emissions from deforestation and degradation, support the sustainable management of standing forests and maintain forest carbon inventories (UNFCCC 2013).

Similarly, as a collaborative initiative of the UN in the face of the continuing disappearance of tropical forests at a worrying rate, it officially launched the UN-REDD Program as a program aimed at

[15] According to Article 3 of the UNFCCC, the PRCD recognizes responsibilities for the problem of climate change. Responsibility is defined by the degree of participation of countries in the increase in global temperature and the capacity of each country to address the problem (UNFCCC 2011).

reducing GHGs from deforestation and forest degradation in developing countries. The UN-REDD Program includes conservation strategies, sustainable forest management and seeks to strengthen the role of indigenous peoples, local communities and other forest-dependent populations and the participation of civil society (Herrán 2012).

The Reducing Emissions from Deforestation and Forest Degradation (REDD) mechanism intervenes when CO_2 is emitted and when forests are damaged or destroyed. The starting point for implementing REDD was for those countries with high rates of deforestation and the potential to reduce them. However, there are many countries that historically have had low rates of forest loss and have retained a large percentage of their forest cover. Given this premise, for political and technical reasons, the REDD concept was expanded to REDD+ (Chacón-Cascante, et al. 2011).

In this context, controversy arises around international cooperative efforts to reduce deforestation in developing countries. REDD+ being an international mechanism that encourages voluntary efforts by these countries to reduce GHGs and increase forest carbon stocks, the UNFCCC designed an International System for determining Emission Reductions and Eliminations (ERs) at the national and subnational levels. Given this, notes that the international framework competes with national emissions trading systems and national REDD+ legislation in the voluntary carbon market (Streck 2020).

3. REDD and REDD+ Mechanism

In the beginning, the issue of carbon emissions from the forestry sector was designated as "avoided deforestation" and not "avoided deforestation", which marked the starting point of the problem. Avoided deforestation only means quantifying achievements in reducing deforestation, even if deforestation continues. The problem becomes more complex when financial compensation is offered for areas where deforestation has been avoided. The optimum would be to avoid deforestation in all countries (Carrere 2012).

In this framework, climate negotiations pushed mainly by developed country governments and corporations and for the purposes of their objectives within the carbon market continued, seeking supposed solutions to the discontent that REDD had already provoked. Negotiations increasingly controlled by corporate power were not concerned with protecting forests and their biodiversity, eradicating poverty, let alone respecting the rights of peoples, indigenous communities and other forest-dependent populations. Deforestation and GHG emissions continued, developed countries continued to use the mechanism as an option that allowed them to pay for polluting, while developing countries saw REDD as a possibility to obtain financial resources for the conservation of their forests (Carrere 2012).

One of the pillars of the mechanism was the PES as the initial REDD scheme of payments conditioned to the reduction of GHG emissions and later translated into PPR (Angelsen 2017a). From the beginning, the PES process meant commodification, subjugation and enslavement of nature to the logic of capitalism. The trade in environmental services fostered impunity because instead of prohibiting deforestation and forest destruction, they compensate for it and avoid combating the problems, undermining the real solutions to the climate crisis with a distraction to changes in modes of production and consumption (Olca 2015).

At this juncture, in 2014 in Peru, more than 150 organizations from around the world expressed their opposition against REDD and extractive industries, to stop capitalism and defend life and their territories. The slogan of controlling deforestation was running out of support, the increasing forest destruction was accepted and even promoted under the slogan of compensation, local participation, improved forest management, development of local populations and even land rights. Peoples and communities were feeling the consequences not only because of intense droughts, floods and other environmental impacts, but also because of the dispossession and plundering of their territories, a consequence of extraction legitimized by the effects of the expansion of the carbon market (Olca 2015).

From the official institutional framework, in response to the growing concern about land use, particularly forest loss and its direct impact on climate change, the Parties to the UNFCCC recognized the importance of advancing the climate change agreement to include a mechanism to reduce emissions from deforestation and degradation. The REDD proposal gained further depth by including conservation, sustainable management and enhancement of forest carbon stocks (UNEP 2018b).

The REDD mechanism soon became REDD+, no longer speaking only of avoiding forest destruction but also of "conservation, sustainable management and enhancement of forest carbon stocks". By focusing on carbon sequestration, the possibility of replacing primary forest with new monoculture plantations or a few fast-growing species was left open; in other words, the door was opened to industrial logging and the monoculture forest industry as future sources of carbon revenue. Carbon was targeted as a climate change mitigation strategy that proposes new and multiple objectives that encompass diverse interests with a large number of different actors and agendas. Therefore, REDD+ is the same REDD only bigger and with the capacity to cause more damage (Kill 2014).

Thus, REDD+ with its focus on reducing carbon emissions has diverted attention from the direct[16] and underlying causes[17] of deforestation to camouflage the violation of the rights of communities in land tenure, the traditional use of their lands and encouraged industrial agriculture, forest monoculture, extraction of minerals, gas and oil and large-scale infrastructure works justifying the development model associated with increasing consumption. The REDD+ initiative, which on account of reducing forest destruction, does not address the real causes, exposes interests that are outside the limits of the mechanism, the real

[16] *Direct causes:* clearing for expansion of different types of subsistence and commercial agriculture, mining, infrastructure development, urban sprawl (UNEP 2018a).

[17] *Underlying causes:* the incidence of population growth, national policies in favor of non-forest land uses, poor governance, tax incentives and subsidies, behavior of national governance preferentially for agricultural commodities, movements of landless farmers who are very poor, resorting to subsistence farming and food insecurity (UNEP 2018a)

perpetrators of the destruction remain untouched as well as their emissions (Sunderlin, Ekaputri, et al. 2014).
To the extent that REDD+ is linked to carbon markets, a currency must be created to meet regulatory or voluntary mitigation obligations. The principle that characterizes carbon markets is the allocation of a right to pollute, which defines the functioning of a cap-and-trade system. When the system recognizes offset credits, these credits become rights to pollute equivalent to the emission rights allocated to the cap-and-trade system (Streck 2020). The major concern is the commoditization of pollution that puts the richest countries and actors at an advantage and creates a global abuse.

REDD+ calls for developed countries to finance developing countries to stop forest destruction in exchange for credit for emissions that may not have occurred. The mechanism requires complex calculations that will determine the amount of carbon stored in a forest, which is difficult to verify. When the figures have been obtained, they are converted into CO_2 equivalent units (the currency of the carbon market), priced and traded on the market as carbon credits. The countries and/or companies that buy, count them as an item that represents part of the compliance with the GHG reduction goals (Fernandez 2015), a convenient situation for the financiers but to the detriment of the indigenous peoples and communities that are weakened in their local food systems by being repressed in their traditional agricultural practices, in their subsistence activities and in the restriction of access to land and forests. Meanwhile, the major drivers of deforestation such as industrial mega-projects, mining, infrastructure and especially industrial tree plantations such as oil palm and soybean, large-scale animal breeding, continue their activities without restrictions; strengthening the industrial agri-food system controlled by corporations that cause climate change (Naranjo 2018).

In this context, it can be observed that over the course of the decade, the basic idea of REDD+ has been to commercialize the carbon stored in the forests as an incentive to reduce GHG emissions. This leads one to think that the mechanism was conceived in such a way that the more deforestation and threats to forests exist, the more REDD+ projects are justified and implemented without any of them focusing on the main objective that defends the creation of the mechanism; so much so that the projects show that they are a threat to the population that lives and depends on the forest. In this way, not only carbon credits are generated to expand and legitimize the activities of the same deforestation actors but also lucrative markets for financial speculation are created (Angelsen, Hermansen, et al. 2019a).

4. Climate finance

A multi-institutional financing architecture has been consolidated at the global level for the implementation of climate finance for transformational change, particularly in difficult areas to mitigate, adapt and reduce vulnerability. As climate finance is key to the Parties to the UNFCCC, the Standing Committee on Finance (SCF) was established in 2010 to track long-term financial resources for low-carbon and climate-resilient development in low-carbon and climate-resilient countries. One of the functions of the SCF is to assist the COP in mediating, reporting and verifying support to developing countries (OECD 2015).

Climate finance aims to "reduce emissions, enhance sinks of greenhouse gases and reduce the vulnerability of human and ecological systems to the negative impacts of climate change and to maintain and increase their resilience" (UNFCCC 2014). To facilitate the provision of climate finance, the UNFCCC outlined financial mechanisms to provide economic resources to developing countries that are in the service of the PA. Adaptation and climate finance are closely related to the objectives of the PA, for this reason, climate change has a broader approach, not only as a phenomenon of temperature increase and GHG emissions but also as an economic and social problem that engages everyone (UNFCCC 2021).

Financing is provided at the local, national or transnational level. It comes from public, private and international cooperation sources, is necessary to mitigate and adapt, protect and restore natural ecosystems, requires large-scale investments to significantly reduce GHG emissions and adverse climate impacts (UNFCCC 2021). It is channeled through an array of avenues such as multilateral funds, bilateral development assistance institutions geared towards climate change; in turn, several developing countries have created and/or adapted national institutions and channels to receive such funding. While the multitude of channels has resulted in increased options and possibilities for recipient countries to access climate finance, it has also become an intricate process in terms of tracking all financing funds (Bird, Watson, and Schalatek 2017).

Over the last decade, climate finance has steadily increased . For the 2015-2016 biennium it reached USD 463 billion; for the 2017-2018 biennium the amount rose to USD 574 billion - an increase of 24 % and for the 2019-2020 biennium it increased to USD 632 billion - a 10 % increase. This allows us to observe that flows have slowed down in recent years, despite the fact that the internationally agreed climate targets for 2030 call for increased climate finance by at least 7 times. Three quarters of global climate investment is concentrated in East Asia and the Pacific, Western Europe and North America, while the other regions received less than a quarter.

East Asia and the Pacific account for approximately 50 % of global climate investments with USD 292 billion in the 2019-2020 biennium and an increase of USD 43 billion compared to the 2017-2018 biennium. It is estimated that more than 80 % of the East Asia and Pacific region was concentrated in China (Buchner, et al. 2021).

Climate finance flows do not meet the estimated needs to achieve the transition to a sustainable zero net emissions and resilient world in this decade, climate investment must increase. Climate finance commitments must be translated into real actions through public and private actors that align with PA objectives (Buchner, et al. 2021).

For a better functioning of climate finance, there are several international entities, institutions and financial mechanisms such as the Global Environment Facility (GEF), the Green Climate Fund (GCF), the United Nations Adaptation Fund (AF), the Special Climate Change Fund (SCCF), the Least Developed Countries Fund (LDCF) oriented to

GHG reduction. In turn, developing countries have increased their public spending on climate change activities through national budgets (UNDP 2012).

4.1. Scope and obstacles of climate finance

Climate finance is the cornerstone for the materialization of agreements, commitments and compliance with climate goals. They are investments that help increase resilience and reduce vulnerability to the impacts of the climate crisis, facilitate the change of production patterns in the agricultural sector and the management of climate disasters. The UNFCCC sets out agreements that must be complied with to build confidence among Parties, and the PA stipulates that all financial flows must be consistent with low-carbon and climate-resilient development. In other words, financial flows must come not only from developed countries but also from developing countries that also have their share of responsibility for environmental pollution (Guzman 2021).

International financial mechanisms should provide clear and coherent guidelines on climate guidelines such as process planning, project implementation, fund management up to the final stage; these guidelines should be followed by the national agencies responsible for implementing climate policy in each country. In turn, governments must ensure that financial funds are used to strengthen environmental activities such as monitoring and the capacity to generate real information at the national level (Sweeney, et al. 2011).

One of the objectives of the PA revolves around the Transparency Framework for action and support (Art. 1316), in this sense all countries must have a MTDS that facilitates reporting the results of climate actions and the support received for this purpose (Samaniego, et al. 2019). The development of a universal methodology would make it possible to know the amount of resources that are pledged, allocated, transferred and, above all, how the resources are used and the information that each country should generate. The lack of a single methodology has shown that climate finance is not consistent between donor and recipient countries because there are less developed countries whose economic capacities are more restricted and others that are on a different development and economic growth trajectory such as China, Brazil and India that should not receive funds from international cooperation (Guzmán, Castillo and Moncada 2017).[18]

Financial transparency is essential in climate action to track and analyze investments. If countries are not transparent about their climate contributions, it is not possible to move from planning to action and thus make progress against climate change. When large amounts of economic resources are invested, there is a high level of complexity and lack of clarity that accompanies many climate processes (Guzman 2021). It is worth asking whether there is transparency in the use of economic resources received by developing countries from multilateral financing funds and whether they are channeled towards priority activities and sectors. In other words, is such financing reaching where

---

[18] "In order to build mutual confidence and promote effective implementation, a strengthened transparency framework for action and support is hereby established, with flexibility to take into account the different capacities of the Parties and based on collective experience" (United Nations 2015).

it should, is it generating the expected impact and is it facilitating the necessary transformation in the territory?

From another perspective, climate finance, despite its importance in meeting national goals, few countries have climate finance strategies as State policies, a situation that is not only due to a lack of resources but also to the political will to address the problem and implement climate actions. Climate finance is a thermometer that measures the priorities of countries and their commitment beyond the institutional framework, therefore, the commitment to climate change is an opportunity to reformulate not only the development model but also the institutional framework (Nemirovsky 2019).

Until recently, climate change was conceived as an environmental issue, but today it is recognized as a problem economic and of great social impact that has not yet been internalized in practice. Developing countries have not yet elaborated their national and sectoral development plans with a climate change perspective. In recent years, these countries have made climate commitments to confront the threat looming over the world; however, these efforts run the risk of being delayed by the development model that promotes oil activities, the expansion of the agricultural, livestock and mining frontiers and their expansion into forested frontier areas with the consequent deforestation. Each of them attracts funds and involves industrial and commercial activities (Salvador 2021).

While climate finance helps to address the effects of climate change, it is also true that the negative impacts are becoming more frequent and intense; therefore, addressing climate change will become increasingly onerous as long as countries and industries continue to delay GHG reductions. Although finance is available, the challenge is that the volume of finance will not be sufficient for the needs of developing countries. It is urgent to rethink climate finance as an integral part of a new model of production and global insertion, which translates into a call to put an end to a linkage model focused on obtaining short-term profits that disregards social and environmental costs, which implies generating value from another perspective focused on non-timber products, sustainable agriculture and reducing or at least not expanding extractive activities (Cabral and Bowling 2014).

4.2. Financial Funds under the UNFCCC

Few countries provide more than 80% of international funding: Norway, Germany, the United Kingdom, Australia and the United States (Olesen, et al. 2018). Multilateral funds managed by the WB, GCF, GEF and the UN-REDD program are responsible for the distribution of approximately one third of international public funding (Norman and Nakhooda 2015). The funds are used to cover transaction, information, implementation, enforcement and monitoring and direct costs of REDD+ activities (Vatn and Vedeld 2011). For the purpose of this research, the actions carried out by the GCF and GEF funds will be evaluated.

4.2.1. Green Climate Fund (GCF)

The GCF, created in 2010, is an operational body of the financial mechanism of the UNFCCC, considered a key source of global support and one of the largest aid funds for developing countries. It is a legally independent entity, whose purpose is to reduce GHGs, improve resilience to climate change and consolidate the objectives of the PA. It began mobilizing financial funds in 2014, captured commitments from 43 contributing countries for USD 10.3 billion mainly from developed countries and other regions. It acquired funding decision-making authority in mid-2015, making it the largest multilateral climate fund with the potential to direct even larger amounts. Its activities, aligned with the priorities of developing countries, especially the most vulnerable, are carried out through the National Ownership Principle (NOP)[19] , for which it established a modality of direct access to financing and thus avoids the presence of international intermediaries. Its actions are based on the use of public investment to stimulate private investment and manage its portfolio of projects through partner organizations known as Accredited Entities (GCF 2019).

Developing countries that are party to the UNFCCC can access support from the GCF to develop REDD+ activities through the Project Readiness Preparation Program and Regular Project Cycle Financing, which will enable them to receive PPPs and carry out their actions to combat climate change. Although funding is focused on GHG reduction, resources are also targeted for adaptation including

[19] "Country ownership and a country focus are the main keys to the fund and agreed that recipient countries should designate a national designated authority (NDA) or focal point. The NDAs will be in charge of program oversight, implementation of the no-objection procedure, ensuring coherence and consistency in all funding proposals, and acting as focal points for communication of the Fund" (Sean 2013).

increasing ecosystem resilience, improving the livelihoods of communities in the most vulnerable regions, and improving food and water security (GCF 2018a).

For the PPR distribution target in 2017, a pilot scheme was launched through which USD 500 million was allocated to countries that meet the requirements of the MTDS implemented by the UNFCCC to reduce GHG emissions from land use and land-use change. Currently four countries have received payments for the equivalent of USD 229 million based on REDD+ PPRs, for emissions reductions of approximately 45 MtCO2eq. Brazil with USD 96.5 million, Chile with USD 63.6 million, Paraguay with USD 50 million and Ecuador with USD 18.6 million for about 3.6 MtCO2eq avoided during 2014 (UNFCCC 2021).

More than a decade has passed since the GCF has been investing billions in public climate finance but the rate of deforestation remains high, especially in the countries that have received the most funding. For example, in Brazil, in the 2014-2015 biennium the government established that environmental conservation outcomes will be reversed if there is no continued payment from the GCF. This meant that the fund had to establish an indefinite financial commitment with the country to maintain its standing forests that in the end were burned in 2019, areas intentionally deforested to end up in agricultural expansion processes; they were resources destined in vain and not in real solutions that caused great destruction for communities, ecosystems, biodiversity and the planet (Lovera-Bilderbeek 2019). Similarly, Colombia and Indonesia in 2020 received USD 130 million from the GCF as PPR but indigenous peoples and communities continue to be ignored and at high risk of deforestation in other areas as lands are cleared for the expansion of palm oil, mining and extraction of other raw materials (Biondi 2020).

The GCF should reject requests for REDD+ funding by PPRs in reference to previous years' deforestation reductions; it should not reward governments that continue to engage in and promote large-scale deforestation because it would be ignoring the increasing rate of deforestation in the countries it funds. For example, the governments of Colombia and Indonesia continue to hand out oil and mining concessions and provide incentives to private industries and agribusiness that destroy forests by implementing infrastructure that these industries require (Lohmann 2020).

4.2.2. Global Environment Facility (GEF)

The GEF is a partnership for international cooperation and was established in 1991 to address accelerating environmental problems as the pilot scheme of the World Bank. Since then it has become one of the largest funders of projects in biodiversity, climate change, international waters, land degradation, ozone layer depletion, persistent organic pollutants, chemicals and waste. Comprised of a global constituency of 183 countries, NGOs, international institutions and the private sector, it has provided more than US$21 billion in grants, channeled more than 5,000 projects in 170 countries for US$114 billion, and supported 133 countries through targeted grants to more than 25,000 civil society and community initiatives (GEF 2020).

The allocation of resources is done with multiple approaches, one of them being climate change based on resources used on results, making sure that developing countries receive their share of funding. From 2014 to 2018, 30 donor countries have committed USD 4.43 billion for REDD+ activities. By 2017, climate change projects accounted for approximately 29% of the cumulative funds which equates to USD 4.7 billion defining it as a major multilateral source of funding for climate change related activities (Bird, Watson and Schalatek 2017).

One of the postulates of the GEF is to direct its investments to promote a transformative change in the systems that generate environmental losses such as energy and food (GEF 2020). However, the administrative decisions of project coordination are organized without taking into account the social relations and realities of the territory where they are applied; "sustainable" production practices are concentrated in territories of productive landscapes where they do not promote changes in the dynamics of the production model that is being developed (Ordenavía, Bernal and Narváez 2021).

The GEF, as an international political actor dedicated to environmental financing, focuses on biodiversity conservation based on an economic logic and designs its strategies according to its interests, which may or may not coincide with the environmental priorities of the countries in which it participates. It elaborates its strategies through a valuation of nature as a "provider of ecosystem services " that is why its major funding is oriented towards biodiversity (Lorenzo 2016). The direction of the GEF has been under the control of the WB and the directive of the donor countries, with strong information restrictions, this connotation has had a great impact on the projects where the fund intervenes externalizing its economic interests and subsidizing the corporations responsible for pollution (Overbeek 2020).

4.3. Lessons Learned from REDD+ Finance

Since 2008, USD 4 billion has been committed to multilateral climate funds to support REDD+ initiatives. Despite the strong interest in market-based mechanisms to invest, the future is uncertain. Since the same year, cumulatively USD 2.4 billion has been allocated for REDD+ activities, of which USD 260 million was approved in 2018. Multilateral climate funds continuously finance projects that support mitigation and adaptation, the GCF has approved at least five projects related to forests and land use. Changes have been observed in the structure of REDD+ and increasing efforts for developing countries to adopt GHG reduction programs with PPR. Multilateral funding for Latin America is recorded at USD 1.2 billion which is equivalent to 52 %, about one third of the total approved funding is concentrated in Brazil with 34 % (Watson and Schalatek 2019).

Norway is the largest contributor of funding to multilateral funds for REDD+ activities whose contribution represents 57% of the total amount committed (Watson and Schalatek 2019). Norway has contributed to climate change mitigation, has invested in over 40 countries and influential institutions such as the WB, UNDP, Food and Agriculture Organization of the United Nations (FAO), conservation groups, and forestry research centers; its investment allowed it to buy support from these sectors for the program and ensure that REDD+ was included in the PA text in 2015 (Lovera- Bilderbeek 2019).

Forest owners and users represent a multiplicity of actors, actions and interests that in the process have been conceptualized as a homogeneous group, but the reality is different. The same connotation has been given to donors who also respond to a variety of interests. The financial contribution of REDD+ is PPR, a complex phase to implement and whose challenges are summarized in who to pay (governments, companies, indigenous peoples, local communities), under what conditions, in exchange for what and how to establish reference levels. In this regard, there is no defined methodology to estimate the possible results, which has led to different understandings or interpretations depending on monetary, political and social interests, causing uncertainty due to biased information and manipulation for their own benefit (Angelsen, Hermansen, et al. 2019a).

In addition, political factors hinder the operation of PPRs, which have led to varying interpretations of what the results are. While payments are based on proven emission reductions achieved in the past, beneficiary countries might translate them into a reward for their efforts, while donors expect these financial resources to be reinvested in strategies to reduce GHG emissions in the future (Angelsen, Hermansen, et al. 2019a).

It is worth remembering that the alleged reduction in GHG emissions is "the result of comparing the actual deforestation rate with the baseline, derived from hypothetical projections of future deforestation or averages of past emissions during periods of maximum deforestation" (Lohmann 2020). Reporting of GHG reductions is likely to remain only in reports, governments may report a period in which they claim reductions have occurred and establish Forest Reference Emission Levels (FRELs) against which they compare deforestation, which may result in fraudulent calculations favorable to the country to obtain PPRs (Lohmann 2020).

The experience of more than a decade has shown that the policies adopted have not been aimed exclusively at carrying out PPR, resources will be provided if the forest is destroyed. REDD+ turned out to be a more abstract mechanism that, using abstract financing tools, mobilizes funds for activities that sequester or conserve carbon emissions in forest and agricultural systems and promote the trade of ecosystem services, thus becoming a promoter of the inclusion of forests in carbon markets (Kill 2014).

The effectiveness of the mechanism has been the subject of controversy because there is no way to guarantee forest conservation on a permanent basis (Lovera- Bilderbeek 2019). The central argument is to keep forests standing that are worth more than felled and that payments are made through governments, NGOs and/or companies that are responsible for deforestation and the main beneficiaries of receiving the funds earmarked to prevent environmental damage. Ideally, these funds should go to the indigenous peoples and communities that conserve the forest that is already conserved and whose governance rights are not recognized. Decrease the millionaire spending on consultancies that prepare methodologies, on entrepreneurs and conservation NGOs that apply intricate REDD+ plans, on pilot initiatives and model projects; while others are in charge of certifying the applications of the first consultants (Angelsen, Hermansen, et al. 2019a).

5. REDD+ Mechanism Ecuador

Ecuador is one of the megadiverse countries recognized worldwide. Thanks to its location and geographical regions, it is a natural laboratory for research and supplier of food, medicines and raw materials; it concentrates part of the world's rich biodiversity and is home to countless species of flora and fauna. With a total area of 256,370 km2 which includes the continental and insular surface. Continental Ecuador, due to its geographic location within the Andes Mountains, is divided by a double mountain range that defines three different regions: Coast, Highlands and Amazon; each with a great variety of climates, soils, landscapes and biodiversity. The continental region is covered by different types of forests whose characteristics depend on climate and soil type. The insular region has an extension of 799,540 ha. of land area, has a large number of animals and variety of endemic plants that depend on the proximity of the sea, it is characterized by having unique species both terrestrial and marine (Ulloa and Sierra 2019).

Likewise, Ecuador's natural conditions vary according to space, elevation, environmental conditions and general characteristics of its ecosystems. The diversity of its arboreal richness is wide and the human impact in each of its regions has been severe. For example, it is estimated that more than 60% of the area of the Coastal Deciduous Forest has been destroyed by human activity, mainly agriculture and cattle ranching. In the Chocó Tropical Rainforest, anthropogenic degradation is high, with almost 75 % of the forest destroyed, it is the region most threatened by deforestation with 74.1 % of damage. The Western Piemontano Forest, in the western foothills of the Andes,

52.1% has been deforested. The Western Montane Forest located between the Mira river basin and the Chanchán and Chimbo river basins, almost half of its area has also been deforested (Ron 2020). In response to this problem, Ecuador has carried out several activities in forest conservation and GHG reduction through instruments such as the PSB, the National Mitigation Plan, which is part of the ENCC, and others.

5.1. Brief description of the Socio Bosque Program

While the REDD+ concept was being developed and negotiated at the UNFCCC, Ecuador was already making innovative efforts to maintain forest cover through the PSB, currently a project, with the participation of several actors, the objective is to conserve the ecological, economic and cultural value of native forests, reduce GHGs, provide financial resources to the impoverished population in rural areas and maintain other ecosystem services apart from carbon capture and storage. Under this scheme the government prioritized areas to sign conservation agreements (Guedez and Guay 2018). The PSB focused on ensuring that the forest maintains its original character and ecological functions, which placed Ecuador as one of the pioneer countries that implemented payment initiatives for forest conservation in its entirety (Flores Aguilar, et al. 2018).

The PSB was developed because of the relevance of native forests. Eighty-eight percent of the 1.3 Mha. incorporated into the program are owned by the participating peoples, indigenous communities and other collective groups. Individual partners were included to increase the area under conservation; however, the lack of clarification of land tenure hindered the long-term projection to incorporate more hectares into the conservation program (Ministry of Environment 2020).

As a result of this initiative, Ecuador has protected more than 1.6 Mha. corresponding to 6.3% of the country's surface and equivalent to 10.7% of the remaining area of existing native forests and natural moorlands (UNDP 2019), it has been determined that the PSB has contributed in decreasing the rate of deforestation at the national level (Ministry of Environment 2020).

According to official information from the PSB, in 2009 the country had a forest cover of approximately 13,038,367 ha, which represents 52% of the estimated surface area of 24,836,000 ha, of which 80% is in the Amazon Region, 13% in the Coast and 7% in the Sierra, which includes various types of forest, including: tropical rainforest, montane forest, high altitude Andean forest and dry forest. In 2014, the forested area reached 12,753,387 ha. this is approximately 51% of the country's surface, which means that there has been an average net deforestation of around 47,497 ha/year, significant figures. Of the total, about 40% are existing forests that are within the National System of Protected Areas (SNAP) and the remaining 60% are in the hands of individual landowners, communes and indigenous communities (Ministry of Environment 2019).

In 2017, the total surface of Areas Under Conservation (ABC) of the PSB reached a total of 1,629,678 ha, a relevant surface. This clarification is made because there is an overlap of partner conservation areas with areas protected by the State equivalent to 297,961.28 ha.

Regardless of the overlap, the agreement of the partners that are within protective forests or protected areas, the conservation commitment is affirmed by the incentive they receive (Ministry of Environment 2019).

The even greater problem is that within the protected areas, extractive activities of non-renewable resources such as oil, mining, illegal timber extraction and even the deforestation of mangroves to build shrimp breeding ponds have affected the ecological reserves of mangroves and wildlife, without considering the prohibition established in the Constitution and in the Forestry and Conservation of Natural Areas and Wildlife Law (Moreano 2012). The environmental impacts of extractive activities make any other productive activity impossible in the affected regions, displacing existing peoples and communities, stripping them of their livelihoods, culture and ways of life.

Through the PSB, the country has conserved 1.6 Mha, although it is still not enough. The Ministry of the Environment as the intermediary agency is in breach of the payment agreement, has delays and has even stopped paying, which represents a threat to conservation objectives. Non-compliance could increase illegal logging, land invasion and illegal expansion of the agricultural frontier. The State must fulfill its commitments and solve the organizational and financial sustainability problems that threaten the program (Montaño 2021).

Incentives as a direct strategy for forest conservation place the PSB as the program that pays a monetary incentive (p/ha. semiannually or annually) to forest owners who wish to conserve forests through a "voluntary agreement" and by signing a twenty-year contract designed under legal conditionalities whose early termination is linked to a series of penal, civil and administrative sanctions (Granda and Yánez 2017).

One of the requirements to enroll in the PSB is that communities must possess communal property titles and part of their income must be used to demarcate and protect the enrolled land, includes the placement of signage or boundaries around the property (Hayes, Murtinho and Hendrik 2017). This allowed the PSB to strengthen tenure security and reduce land conflicts, using demarcation or definition of property boundaries as a tool that influenced the formalization of land tenure (Jones, et al. 2020).

The forest-dependent communities and populations that decide to become partners commit to preserve the ecosystem intact, they can extract products for their self-support but not for commercialization, in no case can they clear a portion of forest for planting and they must take care of the protection zones, in practice they become forest rangers of their forests. However, in the contracts there is no prohibition to carry out industrial extractive activities, "if the state finds oil or minerals in a land enrolled in Socio Bosque, it can exploit it without impediments" (Moreano 2014).

Payments to communities for conserving forests have been controversial; conservation is an inherent activity for them, and the result of their work is millions of hectares of standing forest. The payment has distorted the meaning of protecting their livelihoods and has monetarized nature, in exchange, communities have mortgaged their territories and put their food sovereignty at risk. In addition, the economic resources they receive should be invested in

projects that are approved by the State, for example, in control and surveillance tasks, improvement of community infrastructure, boundary marking and productive development, the results of which must be reported periodically. On the other hand, there were agreements that had minimal and in many cases no risk of deforestation, which led to think that without the PSB, the forests would still be conserved without the need of economic resources to encourage them. These results limited the scope of the program with low additionality[20] (Castellano 2010).

Despite the PSB's good intentions, there are diametrically opposed criteria. Indigenous and environmental organizations in the country have expressed their concern about certain clauses on collective rights in the signed agreements. The PSB was translated into a form of taxation for the conservation of forest resources without recognizing the rights of indigenous peoples and communities to sustainably manage these resources according to their needs. The large number of obligations regarding the conservation and protection of ecosystems is in the hands of the communities, while the Ministry of the Environment has no obligation regarding forest conservation and protection but has the power to unilaterally terminate the agreement (Gonzáles 2011).

While it is true that the PSB recognizes the role of indigenous peoples, communities and other forest-dependent populations in the

[20] Additionality defines the risk that carbon emission reductions will occur even without payments (Atmadja and Louis 2012).

conservation of biodiversity and the protection of ecosystem services, it affects collective rights by limiting communities' access to and traditional use of their forests, further threatening their survival. This requires a comprehensive strategy that includes a work plan in various national and international scenarios with the support of entities committed to natural resources, human rights and the health of the planet (Warmikuna 2016).

RESULTS

SHORTCOMINGS OF REDD+ IN TERMS OF EFFECTIVENESS, EFFICIENCY AND EQUITY

This chapter refers to the process that Ecuador has carried out for the implementation of REDD+ mechanisms. It analyzes the management of the projects of the Amazon Integrated Program for Forest Conservation and Sustainable Production (PROAmazon) with climate funds under the UNFCCC. With the information obtained from interviews with academics, environmental specialists and climate finance experts, an evaluation of the use of REDD+ financial funds is prepared and the 3E+ criteria are assessed through the selected variables.

Climate finance in Ecuador

In 2016, Ecuador formalized the REDD+ implementation phase, the Ministry of Environment issued the REDD+ Ecuador Action Plan "Forests for Good Living" 2016-2025 (PA REDD+) as a set of strategic lines to promote climate change mitigation actions. The mechanism finances programs aimed at reducing deforestation through evaluation and monitoring strategies through the Forest Emissions Reference Levels (NREF)[21] , the Monitoring, Reporting and Verification System (SMRV) and the Safeguards Information System (SIS)[22] (Ministerio del

[21] Forest Reference Emission Levels (NREF) is a methodological tool from which the baseline is established where the reduced emissions resulting from REDD+ implementation are accounted for (UNEP 2018b).
[22] The Safeguards Information System (SIS) is the mechanism through which the implementation of social and environmental safeguards on the ground is reported, referred to as addressed and respected (UNEP 2018b).

Ambiente 2019b).

In preparing the REDD+ PA, two sources of financing for environmental projects are highlighted. The first with resources from international cooperation and climate funds directed to the Amazon Integrated Program for Forest Conservation and Sustainable Production (PROAmazonla), Payment for Results (PPR) and the REDD+ Early Movers Program (REM); and the second with fiscal funds for the Socio Bosque Program (PSB), the Amazon Production Transformation Agenda (ATPA), Forest Control and the Restoration Project (Ministry of Environment 2019b).

In 2017, under the initiative of the Ministry of Environment, the Ministry of Agriculture and Livestock (MAG), UNDP and with financing from the GCF and GEF funds, PROAmazonla was implemented as a government initiative whose commitment is to implement environmental actions and policies to reduce deforestation and promote sustainable and integrated management of natural resources. UNDP is responsible for the financial management of the funds' resources due to its experience in administration, technical assistance and project execution. PROAmazonla aims to link national efforts to reduce GHGs from deforestation, the growth of the agricultural and livestock frontier, strengthen mitigation, adaptation and forest protection efforts, reduce poverty levels, and achieve sustainable human development (PROAmazonla 2021a).

PROAmazonia works in the Northern Amazon in 25 landscapes[23] in 8 provinces: five landscapes in Sucumblos, five in Orellana, five in Morona Santiago, four in Pastaza, two in Zamora Chinchipe, two in Loja, one in Napo and one in El Oro. The prioritization criteria for the definition of these landscapes are related to areas at higher risk of deforestation, areas of importance for the maintenance of water resources and biodiversity (connectivity), and areas of importance for poverty reduction and diversification of the rural economy (PROAmazon 2021a).

PROAmazonia was financed with non-reimbursable funds from the GCF and GEF until 2023 and has a total budget of USD 145.8 million. Of these, USD 53.6 million were delivered in cash and USD 92.2 million were placed as counterpart in kind (wages and salaries, payment to PSB beneficiaries, leases, equipment and inputs) (PROAmazonía 2021a).

According to Patricia Serrano, manager of PROAmazon, the program aims to converge the environmental and productive agenda of the country to generate opportunities and promote the full and effective participation of indigenous peoples and communities, women and youth in decision-making processes aimed at sustainability" (P2). Emphasizes that in the financial strategy of the REDD+ PA: "there is a

[23] "Landscape is any part of the territory as perceived by the population, whose character is the result of the action and interaction of natural and/or human factors. Each of the attributes that make up a landscape, can be classified by biotic elements such as vegetation, fauna, land use, relief, water, etc. From the combination of all of these, the landscape is configured" (UNEP 2018).

financial gap to implement the entire plan, PROAmazon implements part of the plan" (P2) but more economic resources are needed, other alternatives should be sought to leverage more climate funds to continue with the plan (P2).

Francisco Moscoso, Technical Specialist in Monitoring and Follow-up of PROAmazon, adds that the funds obtained from the committed budgets, both those financed by international cooperation and the budget execution with fiscal resources for the period 2020-2025 have determined that: "the funding gap is approximately USD 333 million, which represents more than half of the total cost of the REDD+ PA, which is approximately USD 670 million for that period" (P1). On the NDC side, the conditional scenario is the plan, therefore it is necessary to manage funding by any means to meet the goals set: "it is necessary to work on economic sustainability to continue with the actions, this is one of the reasons why local governments are being included in the implementation of REDD+" (P1). Such is the case of the provincial government of Pastaza, which has developed its provincial implementation plan anchored to the REDD+ PA and has joined the climate working group (P1).

2. Funds under the UNFCCC

The GCF contribution is aimed at co-financing the REDD+ PA as a set of strategic lines to promote climate change mitigation actions. With approximately 26% of the budget, it helps ensure that financial instruments are aligned with the objectives of the REDD+ PA and control agricultural expansion in forest areas (GCF 2020).

The funds are earmarked for the project "Promotion of financial instruments and land use planning for reducing emissions and deforestation", which has an approved amount of USD 41.2 million. The project financing plan is shown in Table 2. The GCF investment covers 16.4% of the project's financial needs (GCF 2020).

Plan de Financiamiento GCF

FINANCIAMIENTO	MONTO	MONTO TOTAL
GCF Fondo Fiduciario		41.172.739
Cofinanciamiento Total		42.835.908
Ministerio del Ambiente	31.755.550	
Ministerio de Agricultura	8.490.000	
FAO	820.900	
PNUD	1.769.458	
MONTO TOTAL		84.008.647

Table 2. GCF Financing Plan
Source: GCF 2020a
Own elaboration

The GEF fund, through the articulation of intersectoral policies and governmental policies, allocates its investment to the project called "Integrated management of landscapes of multiple use and high conservation value for the sustainable development of the Ecuadorian Amazon region". The project has an approved amount of USD 12.5 million and is implemented through coordinated work between MAATE and MAG, with a timeline from 2017 to 2023. The financing plan for the project financed by the GEF is shown in Table 3 (GEF

2019).

The GEF fund investment for the project is USD 12.5 million plus a parallel contribution of USD 49.3 million from the government, UNDP, NGOs, private sector, academia and the International Development Bank, for a total of USD 61.8 million. Compliance with the co-financing is monitored by UNDP and reported to the GEF (GEF 2019).

Plan de Financiamiento GEF

FINANCIAMIENTO	MONTO	MONTO TOTAL
GEF Fondo Fiduciario		12.462.500
Cofinanciamiento Total		49.338.351
Gobierno	34.347.440	
PNUD	1.000.629	
ONG	3.600.000	
Sector Privado	1.986.008	
Academia	4.453.804	
Banca de Desarrollo Internacional	3.950.470	
MONTO TOTAL		61.800.901

Table 3. GEF Financing Plan
Source: GEF 2019
Own elaboration

As of the year 2020, PROAmazonia has intensified its actions in the territory, overcoming the obstacles caused by the pandemic, in order to reach the established achievements. Initially, the Program's budget was structured by components, but since the year 2022, modifications were made to disaggregate budgets and make interventions visible by provinces.

David Romo, director of the Ethnic Diversity Program of the Universidad San Francisco de Quito USFQ, states that the activities covered by the project require a high financial investment and there is a gap between the budgets programmed, executed and committed: "since the beginning of the project management has been slow, the approval process took a long time from June 2017 to the first quarter of 2018" (A2). In addition, he adds that the Sustainability of the project is intermediate considering the economic situation, the political and social risks of the country: "there is no empowerment of the program by the Ministries, this has been a drawback for the achievement of the results" (A2).

3. Variables of analysis

REDD+ among its objectives seeks to reduce GHG emissions at the lowest possible cost and contribute to sustainable development, with this premise, is it possible to evaluate the mechanism considering these three criteria, is REDD+ achieving the goals of reducing GHG emissions - effectiveness, has this goal been met at a minimum cost - efficiency, what are the consequences in terms of distribution and co-benefits - equity (Angelsen and Wertz-Kanounnikoff 2009). To answer these questions it is necessary to deepen the scope of the 3E+ criteria.

According to Arild Angelsen, Professor of Economics at the Norwegian University of Life Sciences (NMBU): "Few studies analyze the impact of local REDD+ initiatives on forests due to the financial, methodological, political and information challenges and the need to assess the 3E+. Local REDD+ projects and programs often include a combination of interventions through incentives and support measures" (A1). It adds that: "incentives are used to reduce deforestation, while support measures - conditional or unconditional on results - are used to help minimize trade-offs between carbon and welfare outcomes" (A1).

For the evaluation of climate finance for REDD+ Ecuador projects under the 3E+ criteria, the research was based on: 1. deforestation rate to assess effectiveness and efficiency; 2. stakeholder participation; and 3. land tenure to assess equity.

3.1. Deforestation Rate

The results of deforestation and forest regeneration (Table 4), according to historical average data and annual rates show that net deforestation (difference between gross deforestation and regeneration) has decreased in the period 1990-2018; however, in the period 2018-2022 has increased significantly.

Tabla4. Cobertura vegetal y la tasa de deforestación 1990-2022

AÑO	DEFORESTACIÓN BRUTA ANUAL PROMEDIO (ha/Año)	REGENERACIÓN BRUTA ANUAL PROMEDIO (ha/Año)	DEFORESTACIÓN NETA ANUAL PROMEDIO (ha/Año)	TASA ANUAL DE DEFORESTACIÓN BRUTA (%)	TASA ANUAL DE DEFORESTACIÓN NETA (%)
1990-2000	129.943	37.201	92.742	-0.93	-0.65
2000-2008	108.666	30.918	77.748	-0.82	-0.58
2008-2014	97.918	50.421	47.497	-0.77	-0.37
2014-2016	94.353	33.241	61.112	-0.74	-0.48
2016-2018	82.529	24.100	58.429	-0.66	-0.46
2018-2020	91.692	4.158	87.535	-0.75	-0.76
2020-2022	95.570	2.547	93.023	-0.78	-0.76

Table 4. Vegetation cover and rate of deforestation 1990-2022
Source: MAATE 2022
Own elaboration

At the regional level, these figures in terms of total deforested area place the country in fifth place after Brazil, Bolivia, Peru and Colombia. Due to its territorial size, Ecuador loses its forests at a faster rate due to the expansion of the agricultural and livestock frontier, the development of infrastructure, mining and hydrocarbon exploitation, and the extraction of timber resources (Paz Cardona 2022).

In Ecuador more than 50 % of the forests are located in the Central Amazon, North Coast and other tropical areas (Ministerio del Ambiente 2020a). By 2018 the least deforested natural region was the Amazon region with a remnant of about 83 % of the original forest area, about 48 % of the original natural forest area has remained in the Andean region and about 27 % of the original remnant is recorded in the Coast (Sierra, Calva and Guevara 2021).

The main deforestation hotspots are in the Chocó-Darién and the Amazon Basin. The Chocó-Darién, known for being one of the richest areas in biodiversity and for having high rates of endemism, is highly deforested on the Ecuadorian side (Fagua, Baggio and Ramsey 2019). Central Amazonia, in comparison to Chocó-Darién, has a lower rate of deforestation despite the fact that it has also experienced a steady decline in its forests (Ministerio del Ambiente 2020a). The rate of deforestation in some protected areas is also of concern, for example, the Mache-Chindul ecological reserve, located on the Ecuadorian coast, which has lost 39% of its forests. If the current rate of deforestation continues, significant original forest areas will have been lost within thirty years (Paz Cardona 2019).

Only two of the country's six Amazonian provinces account for 46%, corresponding to 287,000 hectares of all deforestation detected between 2001 and 2020. In Morona Santiago more than 25% of the forest was lost, representing 158,000 hectares, and in Sucumbíos nearly 21%, equivalent to 129,000 hectares. Extractive activities such as mining and hydrocarbons are present in both provinces. If the current rate of deforestation continues, significant areas of original forest will have been lost within thirty years (Paz Cardona 2022).

Manuel Shiguango, territorial technician CONFENIAE / UN REDD+, states that one of the concerns of the project funded by GEF is:

> The emphasis that has been placed on transforming the productive sector through sustainable forest management practices, what kind of sustainable practices is it intended to introduce? Is it tree plantations, relying on fast-growing species such as eucalyptus, palm, soybean and monoculture plantations? (C1).

He adds that work must be done to restore the ecological functions of forests: "the missing issue is the need to reduce overconsumption and industrial production of monocultures for export with serious consequences for peoples, communities and forests" (C1). He adds that deforestation and GHG emissions will continue if funding continues to be directed to land use transformation in selected landscapes, causing further damage to peoples, communities and small farmers (C1).

In this context, according to Cristina Garcia, WWF Forest and Water Program Officer, "measuring the effectiveness and efficiency of GCF and GEF funded projects on deforestation rates is complicated" (O2).

The NREF is in the process of being revised, as are the official results of deforestation dynamics for the period 2015-2018. It adds that: "GHG reduction is a complex process that involves political, economic, administrative, technical, social, and other issues; it is not an infrastructure program" (O2). The most equitable and lasting results are those where the local population participates in the design and implementation of the REDD+ program (O2).

Patricia Serrano adds that another factor that has made it difficult to measure the effectiveness and efficiency of the deforestation rate is that the monitoring system is calculated at the national level and not by projects or areas, because the forest monitoring system is national. In response to this: "PROAmazon in coordination with MAATE are preparing an approximate estimate of the annual national deforestation" (P2).

Given the above, it is evident that REDD+ funds are not effective or efficient from whatever source they come from, whether through donations, grants, debt swaps or international cooperation, because they have not been able to reduce deforestation to the expected levels. While it is true, there has been a decrease in the rate of deforestation until 2018; however, with the projects being implemented, the rate of deforestation has increased from 2018 to 2022. REDD+ has not been able to address the problem it was supposed to solve: reducing deforestation, promoting conservation, and the sustainable management and use of the country's forest resources.

3.2. Stakeholder participation

For more than a decade, the country has been working on the preservation of its natural resources, in the orientation and vision of public policy through tools for the management and conservation of forests, to guarantee the right to live well in a healthy and ecologically balanced environment for its peoples and local communities and society in general. To this end, the government at through the MAATE directs its efforts through the National Forestry Directorate to comply with the law, guaranteeing the sustainable management of forests and activities related to the use and commercialization of their natural resources (Ministry of the Environment 2011).

The purpose is to achieve efficient governance that focuses on the maintenance and restoration of environmental goods and services with a view to the conservation of forest ecosystems and biodiversity. Multisectoral dialogue and the continuous dissemination of information are necessary aspects, not only to achieve ownership of the projects and the integration of the different stakeholders, but also to maintain transparency in access to information on all processes (Ministry of the Environment 2011).

Joint construction processes have been generated with the participation of representatives from different sectors, these contributions are embodied in planning and public policy instruments such as the Organic Environmental Code (COA), the National Climate Change Strategy 2012-2015 (ENCC), the National Forest Assessment (ENF), the National Territorial Strategy (ETN) and the REDD+ PA (Ministry of

Environment 2019a).

Environmental governance is complex and is given by the plurality of institutions with different levels of legalization, membership, jurisdictional scope and even with different degrees of relationship between them. Therefore, the complexity of the country's governance also manifests itself in REDD+ making it difficult to quantify and qualify efficiency and equity (Fariborz, Nielsen and Dubber 2019). The institutional framework and governance system are two parameters that determine advantages or limitations in the application of the 3E+ criteria (Kambire et al. 2016). For example, in the face of more institutions and processes, one may have more platforms for inclusion but may increase coordination gaps (Zürn 2018). REDD+ governance is a system of interaction on climate change, biodiversity, forestry and sustainable development; the mechanism provides a nexus where several institutions from different domains collaborate or compete to establish rules, financing, implementation and evaluation systems (Gupta, Pistorius and Vijge 2016).

In this context, the government must readjust its interaction to achieve the planned results. REDD+ environmental policies require national ownership and inclusive political processes through a clear definition of the governance structure in the face of interests that drive deforestation (Wong, Luttrell, et al. 2019). Such factors have clearly not been developed in the country due to lack of clear policies, absence of national ownership, complex political processes that are not rooted in social and environmental issues that have not been developed in terms of vulnerable groups.

Projects under the REDD+ mechanism are channeled considering the institutional framework and forest governance. The challenge of PROAmazonian design is to integrate the organizational work of the two Ministries with the support of UNDP (GEF 2020a). Forest management does not depend on PROAmazon, it is under the control of MAATE and in parallel with other entities and actors working in the process of planning and execution of practices of forest ecosystems that make up the REDD+ Working Table (PROAmazon 2021).

3.2.1. REDD+ Working Table

In 2012, the MoT[24] was created, comprised of 41 organizations (Annex 2) as a national platform for dialogue, involvement, participation, deliberation, consultation and monitoring of key stakeholders and whose function is to monitor the implementation of REDD+ measures and actions in Ecuador. It was institutionalized by the MAATE in the year

2017[25] to ensure that the REDD+ readiness and implementation phases consider the vision and contributions of all stakeholders, both those who have rights to use forests and the agents of direct and underlying causes of deforestation and forest degradation. The scope of the MoT is aimed at the implementation of REDD+ policies and actions, social and environmental safeguards (SSA) and accountability and access to

---

[24] The MoT started 10 years ago. From 2013 to 2015 was the first period, from 2016 to 2019 the second and from 2020 to 2023 the third.

[25] At the time of the research, the MoT was in its third period of operation. It has met 12 times in accordance with the governance model, achieving several inputs that have strengthened REDD+ implementation at the national level (Ministry of Environment 2023).

information on project progress (PROAmazon 2021a).

During the third period of operation of the MoT, it has met 12 times in accordance with the governance model; however, the territories where the projects intervene present problems that need to be resolved, such as the lack of transparency in the information on activities. Community inclusion has not been developed in an effective and equitable manner, and work must be done jointly with PROAmazon (G1).

Alliances between actors involved in the social, technical and political processes should be strengthened, particularly with community leaders, women and youth, aiming at the common good through organizational and territorial articulation spaces. These articulation spaces are key to develop strategies, avoid conflicts and contribute to the promotion of a positive forest management according to the community reality, committed to environmental processes in the medium and long term, projecting sustainability after the closure of PROAmazon (O1).

The territories where the projects intervene present problems that need to be solved: "as long as there is no transparency in the information on the activities of the MoT, it is not possible to make a judgment on the work they carry out" (G1). For example, it is expected that the Environmental Education Plan proposal in the province of Pastaza will motivate participants to become aware of forestry education that is oriented towards the common good, protection and good use of forest resources. There is a need to go deeper into this topic, community inclusion has not been developed in an effective and equitable manner, and it is necessary to work together with PROAmazonia (G1).

On the other hand, PROAmazoma has the support of MAATE, MAG and local governments that are strategically aligned with the vision of the projects on sustainable production issues; however, they neglect the participation of the people and communities. It has been observed that: "much of the work is concentrated on the administrative part, while the essential, technical and strategic issues that should have more attention are little attended to" (O1). It also considers that PROAmazoma has a great challenge to solve the SSA, the benefit distribution plans and the participatory processes that if they are not fulfilled could reflect a weakening of the social objectives of the program (O1).

Likewise, Jaime Toro mentions that "within PROAmazonia's organizational structure, management at the provincial level shows evidence of weak and concentrated governance" (O3). Environmental, social and technical factors should be evaluated and integrated, as well as focusing on local needs and redesigning more solid and equitable forestry plans. UNDP's role focuses more on administrative issues and less on technical issues, leaving aside nature and indigenous peoples and communities. He adds that the Ministries in charge should work in coordination in order to strengthen the work of the program to ensure the quality of management and therefore the achievement of results. In addition: "there are other structural limitations, such as the short periods for project execution, the lack of a methodology designed by the mechanism itself, the bad hierarchical habit of not being accountable, etc." (O3). It is suggested that the teams in charge of the Ministries should be agile in the review of the technical aspects by so that the reporting of the reports is timely and the execution of the activities takes less time than expected (O3).

It appears that the results of the MoT meetings ratify that its members are working in coordination with the activities carried out through REDD+ projects. An action plan must be drawn up to halt the destruction caused by deforestation following the expansion of industrial agricultural monoculture plantations, industrial cattle ranching in forests, cash crops and other activities supported by global food corporations through the linkage to certification standards promoted by REDD+ (O1).

3.2.1. Indigenous peoples and communities

Ecuador has indigenous peoples and communities within its population, with a greater presence in the Amazon region and in the highlands. The term Nationality is defined as a "group of millenary peoples that preceded and constituted the Ecuadorian State, that define themselves as such, that have a common historical identity, language and culture, that live in a determined territory through their institutions and traditional forms of social, economic, legal, political organization and exercise of authority" (INEC 2006). Indigenous peoples are defined as "original collectivities, made up of communities or centers with cultural identities that distinguish them from other sectors of Ecuadorian society, governed by their own systems of social, economic, political and legal organization" (INEC 2006).

Each indigenous people and nationality express their worldview in close relation to their habitat, forests and natural resources. In the wording of REDD+ policies, the concept of communities as beneficiaries of the mechanism and agents for its implementation is generally emphasized; therefore, the measures and actions of the mechanism consider the cultural values, ancestral knowledge and traditional productive activities of communities, peoples and nationalities (Ministry of the Environment 2016). Under this scheme, the REDD+ PA is built within a process of dialogue and participation of national, provincial, cantonal and local *stakeholders* considering the environmental and cultural diversity of the country (Ministry of Environment 2019c).

According to David Yedra, sustainable productive development requires strengthening and improving the communities' production chain, which requires training in different production areas such as the use of fertilizers, harvesting times and having the necessary raw materials and inputs until reaching the final product and marketing; this is the production chain in which the projects must work because these processes demand financial resources (G1).

Patricia Serrano adds that PROAmazonía has signed an agreement with the Confederation of Indigenous Nationalities of the Ecuadorian Amazon (CONFENIAE) for the implementation of the projects through the involvement of the communities, the calls are made through CONFENIAE as the official channel for participation: "currently the projects are working on five life plans for the communities that are translated into their language, with the purpose of generating a participatory process in compliance with the SSA, prioritizing the participation of women" (P2).

The implementation of REDD+ projects has generated divided opinions regarding the rights of indigenous communities. Although the projects have made some progress through technical capacity building in the GADs, they have allowed indigenous and community organizations to make their presence evident as participatory actors in the deforestation and conservation processes through the SSAs; however, the implementation of the projects has not been efficient or equitable in the planning and optimization of funds to support conservation, restoration and sustainable production in appropriate areas of community lands. There is a lack of agility in the coordination of actions by the Ministries in charge of directing the funds to make transparent the rights of the communities with effective participation and consultation in the regularization of their land rights, among others. The indigenous representations consider that most of the financial resources have been channeled to the GADs, NGOs and consultants, which has prevented the benefits from reaching the indigenous peoples and communities as forest protection actors (Aldea 2019).

The participation of indigenous peoples and communities due to lack of knowledge about the REDD+ SSA, has had limitations that must be overcome through participatory processes that involve their presence through dialogues, easily disseminated didactic material according to their language, which articulates and strengthens community organizations within their worldview and interaction with nature that requires coordination, involvement, continuous monitoring and a team of responsible technicians who take ownership of the process (Suarez 2017).

REDD+ created expectations among indigenous peoples and communities by announcing that it would combat the problem of deforestation, improve forest management, guarantee local participation, improve their income and even protect the implementation of territorial rights. However, in practice the mechanism has benefited a group of actors, expanding the logic of the price to nature, reducing the problem of deforestation to the monitoring and commercialization of CO_2, adapting local expectations in an instrumental manner, depoliticizing and camouflaging power relations between the actors involved and weakening the priorities to be resolved such as land tenure and indigenous rights (Vásquez 2013). The mechanism is not an initiative of the communities nor has it mitigated the needs and threats they face, they received promises of benefits and employment but what they received was harassment, restrictions on land use and the inculpation of being responsible for deforestation (Kill 2017).

According to Duval Llaguno: "one of the biggest problems of indigenous peoples and communities is their difficult living conditions, lack of opportunities and economic incentives to help them emerge, which has motivated the need to expand their agricultural frontier in territories that were traditionally destined for conservation" (B1).

REDD+ has been criticized because from the beginning and in the construction process, financial resources have been allocated to define the NREF, SMRV and SSA, neglecting the participation and consultation of indigenous peoples and communities as relevant stakeholders. "The communities have expressed their disagreement with REDD+ management because they are the ones who take care of the forests and do not receive the necessary help to conserve them" (O2).

3.2.2. Land Tenure

Land tenure is important in the planning and implementation of the REDD+ mechanism, as it is the basis on which the distribution of benefits and implementation of projects is built. The REDD+ mechanism promotes investment and forest management in areas that are generally remote from urban centers and in places that are difficult to access. The lack of legal security of tenure is one of the main obstacles to investment in the mechanism. Clarifying and providing security over land tenure rights is the first step in the REDD+ readiness process (FAO 2016).

Tenure management requires a comprehensive public policy and long-term action that includes technical, legal and financial resources, as well as the participation of various actors collaborating in legalization (Hayes, Murtinho and Hendrik 2017). The forestry law does not allow the existence of private property within the protection, forest heritage or protective forest areas that have been declared in documents. Such provision was made without considering the participation of indigenous peoples and communities. For this reason, the overlapping of protected areas and the lack of knowledge by indigenous organizations of some protection areas has led to claims about autonomy for the management of their territories, on the grounds that many of them were inhabited by indigenous peoples and communities before being declared as protected areas (Moreano 2012).

Inequity, illegality and inequality in land tenure is a critical problem in the country and one of the highest in Latin America, considering the size compared to other countries in the region. The Gini coefficient used to measure inequality in access to land resources is 0.81, which is a worrying result (León and Rivera 2020).

Land tenure is a controversial issue. In protected areas, there are ranches that have land titles, which is a problem that needs to be resolved: "the owners of these areas continue to work without considering that they are protected areas, generally in the highlands. A similar situation occurs with areas of the Amazonian communities that have been declared protected and conservation areas, such as Yasuní Park, where oil concessions have been granted" (A2). There is still work to be done that will have to be resolved by the current government (A2).

In addition, land tenure has been linked to other conditioning factors such as oil and mining exploitation, which have changed land use in the Amazon region and are a cause of deforestation and environmental deterioration. Inequity, illegality, insecurity and lack of transparency in land tenure have been conditioning factors of vulnerability, an eminently social phenomenon that affects indigenous peoples and communities and the sustainable development of the country (León and Rivera 2020). According to Patricia Serrano: "land tenure is not an axis of action of PROAmazonía" (P2). In the projects financed with GCF and GEF funds established in the REDD+ PA, the prerequisite for accessing benefits is a land title or that the land has been cleared (P2).

It is evident that in recent years REDD+ projects have not defended, let alone strengthened the rights of peoples, indigenous communities and other forest-dependent populations; on the contrary, they have established new packages of property rights in favor of various powerful actors (Cabello 2014).

Since 2008, three consecutive policies have been created in relation to land: the Plan Haciendas in 2008, the Plan Tierras in the period 2009-2013 and the Plan for Access to Land for Family Producers and Mass Legalization in the Ecuadorian Territory (ATLM) in 2018. However, in the context of social and political instability that has marked the country, its implementation has been ineffective and inequitable due to factors such as: the limited participation of the actors involved, the limited knowledge of the territory, the lack of feasibility studies, the inaccurate calculation of the price per hectare, the scarcity of reliable information on the amount of affected areas and families involved, the lack of evaluation of the payment capacity of the beneficiaries, the inaccuracy of the areas delivered to each family, the restricted access to credit, irrigation, among others. This has hindered progress in the property titling process, which is a work in progress. The reality so far is a high percentage of small plots and a constant struggle of the different cultural identities to materialize their recognized land rights.

There are around 200,000 families in the country that do not have land tenure security and are constantly threatened by the expansion of extractive, oil and mining activities that leave out the sovereignty of indigenous peoples and communities (Ramos, 2022). In addition, the complexity and high monetary cost of regularization protocols make it difficult to obtain land titles. Similarly, land titling in protected areas is a critical problem due to the lack of reliable information procedures, limited physical delimitation systems and the presence of registration mechanisms that cause low effectiveness in their management and enforcement. Resolving land tenure problems is a task that the governments in power must solve, even though it is a complex and costly process, but it is key for the sustainability of ecosystems, biodiversity conservation and the security of indigenous peoples and communities.

CONCLUSIONS

Forest ecosystems participate in the fight against climate change. Ecuador in the last three decades has suffered a constant process of deforestation, the Amanzonia forests may continue to become extinct as has happened in other areas. The loss of forests means for the country to lose thousands of endemic species of flora and fauna unique on the planet with environmental, economic, social and cultural repercussions. Faced with this reality, the country has committed to reduce the levels of deforestation and forest degradation, to promote sustainable production systems to access climate finance.

In this sense, the research sought to evaluate climate finance funds under the UNFCCC for the REDD+ mechanism. Three specific objectives were articulated around this general objective that fed the research work. First, the characteristics of the climate funds that have been invested in the country in the period 2017-2020 were identified. Second, we proceeded to evaluate the effects that GCF and GEF funds have generated in the projects managed by PROAmazonia. To achieve these objectives, information from academic literature and official documentation from MAATE, GCF and GEF was used. Interviews with academics, environmental specialists and climate finance experts were a relevant source for the analysis of the variables under study (deforestation rate, *stakeholder* participation and land tenure) as well as to determine the scope of climate investment in the achievement of the 3E+ criteria. Based on the results found, certain guidelines for the management of public policies on environmental mitigation associated with REDD+ were outlined. With this brief contextualization, the main results of the research are presented.

Once PROAmazonia's activities were completed, the results show that climate finance has not been able to reduce the deforestation rate, which is considered one of the highest in Latin America. The average annual gross deforestation (ha/year) for the 2016-2018 biennium was 82.5 and in the 2020-2022 biennium it was 95.5, i.e., there was an increase from -0.66 % to -0.78 %. This is a consequence of factors such as extractive activities such as oil, mining, timber, intensive agriculture, livestock, among others.

The participation of indigenous peoples and communities as the actors who live in and depend on forests has been underestimated. Ideally, these funds should be allocated to the indigenous peoples and communities that conserve the forest which, although already conserved, their rights are not recognized. The millions of dollars spent on consultancies that prepare methodologies and conservation NGOs that apply intricate REDD+ plans in pilot initiatives and model projects, while others are responsible for certifying the applications of the first consultants, should be reduced. An active participation of indigenous peoples and communities should be generated and dialogues should be built with technicians and implementers, this requires a joint effort to empower and strengthen such activities.

No budget has been allocated within the climate funds to clarify land ownership. The mechanism promotes investment and management of forests in healthy areas that are remote from urban centers. The lack of legal security of tenure is one of the main obstacles to investment;

therefore, clarifying and providing security of tenure rights is the first step in the REDD+ readiness process.

In the management of projects in the country there have been a series of adaptive activities caused by a set of measures where conditionality has hindered the work, which has led to delays and slow progress in the projects.

Finally, the application of climate finance has not achieved the expected results and has been ineffective, inefficient and inequitable. Political instability has had an impact on project intervention areas, has generated slow actions on the part of certain actors, and there are no clear guidelines for executing the required actions within the framework of their competencies and specific lines of work. Although there are certain achievements, there is still much to be done to identify local priorities and interests that are in line with national objectives.

BIBLIOGRAPHY .

Angelsen, Arild, and Arun Agrawal. 2009. *"Using community forest management to achieve REDD+ goals."* Realising REDD+: national strategy and policy options (1): 201-212.

ALDEA. 2019. "REDD+ and indigenous peoples in Latin America." *Latin American Association for Alternative Development.* January 16, 2023. http://www.fundacionaldea.org/noticias-aldea/felr836j7b9g43r5a9edtajnr7hw87.

Angelsen, Arild, Christopher Martius, Veronique De Sy, Amy Duchelle, Anne Larson, and Thu Thuy Pham. 2019. "REDD+ enters its second decade." In *REDD+: Transformation Lessons and New Directions*, by Arild Angelsen, Christopher Martius, Veronique De Sy, Amy E Duchelle, Anne M Larson and Thu Thuy Pham, 1-14. Bogor-Indonesia: CIFOR.

Angelsen, Arild, Erlend Hermansen, Raoni Rajão, and Richard Van der Hoff. 2019a. "Payment for results Who should be paid and in return for what?" In *REDD+: Transformation Lessons and New Directions*, by Arild Angelsen, Christopher Martius, Veronique De Sy, Amy E Duchelle, Anne M Larson and Thu Thuy Pham, 45-60. Bogor-Indonesia: CIFOR.

Angelsen, Arild, Maria Brockhaus, Amy E. Duchelle, Anne M. Larson, Christopher Martius, William D. Sunderlin, Louis V. Verchot, Grace Wong, and Sven Wunder. 2017. "Learning from REDD+: a response to Fletcher et al." *Conservation Biology 31,* (3): 718-20.

Angelsen, Arild, Maria Brockhaus, William Sunderlin, and Lee Verchot. 2013. *"REDD+ Analysis: Challenges and Options."*

Bogor-Indonesia: CIFOR.

Angelsen, Arild, Markku Kanninen, Maria Brockhaus, William D. Sunderlin, Sheila Wertz-Kanounnikoff, and Erin Sills. 2010. "REDD+: From global to national." In *REDD+ Implementation: National Strategy and Policy Options,* by Arild Angelsen, Maria Brockhaus, Sheila Wertz-Kanounnikoff, Erin Sills, William D. Sunderlin and Markku Kanninen. Bogor-Indonesia: CIFOR.

Angelsen, Arild, and Sheila Wertz-Kanounnikoff. 2009. "What are the key issues in REDD design and what are the criteria for evaluating options?" In *Moving Forward with REDD Issues, Options and Implications*, by Arild Angelsen: 11-22. Indonesia: CIFOR.

Angelsen. Arild. *2017a . "REDD+ as result-based aid: general lessons and bilateral agreements of Norway." Review of Development Economics 21 (2): 237-64.*

Atmadja, Stibniati, and Louis Verchot. 2012. "*A review of the state of research, policies and strategies in addressing leakage from reducing emissions from deforestation and forest degradation (REDD+)."* Mitigation and Adaptation Strategies for Global Change 17, no. (3): 311-36.

World Bank. 2021a. *"Gini Index - Ecuador""*. Accessed June 4, 2021. https://www.ecuadorencifras.gob.ec/documentos/web-.

inec/POBREZA/2021/June-2021/202106_PovertyandInequality.pdf (last accessed December 2022).

Barba-Romero, Sergio, and Jean Charles Pomerol. 1997. *"Multicriteria decisions. Fundamentos teóricos y utilización práctica"*. Servicio Publicaciones Universidad de Alcalá. Alcalá de Henares, Spain:

420.

Bayrak, Mucahid Mustafa, and Lawal Mohammed Marafa. 2016. "Ten Years of REDD+: A Critical Review of the Impact of REDD+ on ForestDependent Communities." *Sustainability 8,* (7): 620.

Berruezo, Javier Aldaz, and Julio Díaz. 2017. *"Status of the United Nations Framework Convention on Climate Change.* Summary of the Paris, COP21 and Marrakech, COP22 Summits." *Journal of Environmental Health 17* (1): 34-39.

Biondi, Pedro. 2020. *"New briefing explains the basics of REDD+ and why it's so contentious"."* Global Forest Coalition, (6): 1-8.

Bird, Neil, Charlene Watson, and Liane Schalatek. 2017. *"The global architecture of climate finance. Background information on climate finance."* Climate Funds Update.

Blobel, Daniel, Nils Meyer-Ohlendorf, Carmen Schlosser-Allera, and Penny Steel. 2006. "*United Nations Framework Convention on Climate Change: Handbook.*" United Nations Framework Convention on Climate Change.

Intergovernmental and Legal Affairs of the Climate Change Secretariat.

Buchner, Barbara, Alex Clark, Angela Falconer, Rob Macquarie, Chavi Meattle, and Cooper Wetherbee. 2021. "*Global Landscape of Climate Finance 2021."* Climate Policy Initiative (9): 45-52.

Cabello, Johanna. 2014. "*Masking the destruction: REDD+ in the Peruvian Amazon." World Rainforest Movement, (5): 317.*

Cabral and Bowling, Roberto. 2014. *"Sources of financing for climate change."* ECLAC Financing for Development Series, (254): 36-49.

Clary, E. Gil, and Mark Snyder. 2002. *"Community involvement:*

Opportunities and challenges in socializing adults to participate in society." Journal of Social Issues 58, (3): 581-591.

Carrere, Ricardo. 2012. *"A critical view of REDD"."* Revista Semillas. World Rainforest Movement, (46): 5-18.

Castellano, Eliseo. 2010. *"Sobre REDD+ y el programa Socio Bosque".* Acción Ecológica.

ECLAC. 1989. *"Glossary of Terms Related to External Debt Management."* Economic Commission for Latin America and the Caribbean, (49): 1-17.

Chacón-Cascante, Adriana, Juan Robalino, Brenes Muñoz, and Maryanne Grieg-Gran. 2011. *"Reduced Emissions due to Reduced Deforestation and Forest Degradation (REDD and REDD+)".* Instrument Mixes for Biodiversity Policies. POLICYMIX Report 2, (2): 145-161.

Chhatre, Ashwini, Shikha Lakhanpal, Anne M. Larson, Fred Nelson, Hemant Ojha, and Jagdeesh Rao. 2012. "*Social safeguards and cobenefits in REDD+: a review of the adjacent possible*." Environmental Sustainability 4, (6): 654-60.

UNFCCC. 1992. "*What is the United Nations Framework Convention on Climate Change."* 2012. Accessed December 5, 2020. https://unfccc.int/es/process-and-meetings/the- convention/what-is-the-united-united-nations-framework-convention-on-climate-change.

climatic.

. United Nations Framework Convention on Climate Change. 2013. *"Forest Carbon Partnership Facility (FCPF)."* Accessed

December 2, 2020. https://unfccc.int/sites/default/files/redd_20130228_fcpf_update_sp_ feb_2013_f inal.pdf.

. United Nations Framework Convention on Climate Change. 2014a. "*Warsaw Framework for REDD-plus*." Accessed December 10, 2020. https://unfccc.int/topics/land-.
use/resources/warsaw-framework-for-redd-plus.

. United Nations Framework Convention on Climate Change. 2021. *"UN climate change process steps up action on deforestation."* Accessed April 26. https://unfccc.int/es/news/el-proceso-de-climate-change-of-the.
un-intensifies-action-on-deforestation

Dawson, Neil M., Michael Mason, David Mujasi Mwayafu, Hari Dhungana, Poshendra Satyal, Janet A. Fisher, Mark Zeitoun, and Heike Schroeder. 2018. *"Barriers to equity in REDD+: Deficiencies in national interpretation processes constrain adaptation to context."* Environmental science & policy 88: 1-9.

Dewan, Angela. *"The great challenge of REDD+: Clarifying forest land rights."* Center for International Forestry Research (CIFOR), 2011.

Di Gregorio, Monica, Maria Brockhaus, Tim Cronin, and Efrain Muharrom. 2013. *"REDD+ Implementation: Politics and Power in National REDD+ Policy Processes."* In *REDD+ Analysis: Challenges and Options,* edited by Arild Angelsen, Maria Brockhaus, William Sunderlin, L Verchot. Bogor: CIFOR.

Duveiller, Gregory, Josh Hooker, and Alessandro Cescatti. 2018. "*The mark of vegetation change on Earth's surface energy balance."* Nature communications 9, (1): 1-12.

Ece, Melis, James Murombedzi, and Jesse Ribot. 2017. *"Disempowering democracy: local representation in community and carbon forestry in Africa."* Conservation and Society 15, (4): 357-370.

Fagua, Camilo J., Jacopo A. Baggio, and R. Douglas Ramsey. 2019. *"Drivers of forest cover changes in the Chocó-Darien Global Ecoregion of South America*." Ecosphere 10, (3): 5-38.

FAO. 2016. *"Tenure and REDD+: Developing Enabling Tenure Conditions for REDD+."* UN-REDD Policy Bulletin (6), Rothea, Ann-Kristin & Munro-Faure, Paul. Rome.

. 2015. *"Forest Resource Assessment Work. Terms and Definitions""*. Rome (180).

. 2020. "*The State of the World's Forests. Forests, Biodiversityand People"."* Rome: FAO.

. 2020a. *"Global Forest Resources Assessment 2020. Key findings*." Rome.

. 2003. *"Land tenure and rural development. Studies on land tenure 3""*. Rome-Italy.

. 2003a. "*Forests, the Global Carbon Cycle and Climate Change"*. Proceedings XII World Forestry Congress.

. 2016. "*Tenure and REDD+: Developing Enabling Tenure Conditions for REDD+*." UN-REDD Policy Bulletin (6), Rothea, Ann-Kristin & Munro-Faure, Paul. Rome.

. 2015. "*Forest Resource Assessment Work. Terms and Definitions*."

Rome (180).

Fariborz, Zelli, Tobias Nielsen, and Wilhelm Dubber. 2019. *"Seeing the forest for the trees: identifying discursive convergence and dominance in complex REDD+ governance*." Ecology and Society 24 (1).

Fernandez, Raúl. 2015. "*REDD+ projects and how they undermine peasant agriculture and real solutions to address climate change*." World Rainforest Movement (244).

Fry, Ian. 2008. "*Reducing emissions from deforestation and forest degradation: opportunities and pitfalls in developing a new legal regime.*" Review of European Community & International Environmental Law 17 (2): 166-82.

GCF. 2020. *"Priming Financial and Land Use Planning Instruments to Reduce Emissions from Deforestation."* In *Interim Evaluation,* edited by Javier Jahnsen, Fernanda Salinas and Adriana Bustillo. Quito: GCF.

. 2020a. *"Priming Financial and Land Use Planning Instruments to Reduce Emissions from Deforestation".* In *Interim Evaluation,* edited by Javier Jahnsen, Fernanda Salinas and Adriana Bustillo. Quito: GCF.

. 2019. *"United Nations Development Program COUNTRY: Ecuador PROJECT DOCUMENT."* Integrated Management of Landscapes of Multiple Use and High Conservation Value for the sustainable development of the Ecuadorian Amazon Region.

. 2018a. *"GCF in Brief: REDD+"".* May 5. https://www.greenclimate.fund/sites/default/files/document/gcf-brief- redd_0.pdf.

GEF. 2020a. *"Sustainable development of the Ecuadorian Amazon: integrated management of multiple-use landscapes and high-value conservation forests". Mid-Term Review.*

. 2019. *"United Nations Development Programme COUNTRY: Ecuador PROJECT DOCUMENT"*. Integrated Management of Multiple Use Landscapes and High Conservation Value for the sustainable development of the Ecuadorian Amazon Region.

Gonzáles, Javier Dávalos. 2011. "El convenio del Progama Socio Bosque y las comunidades indígenas en Ecuador". *Amazon Watch* (19).

González, Humberto, Antonio Brenes. 2020. "*The Lorenz curve and the Gini coefficient as measures of income inequality*." REICE: Electronic Journal of Research in Economic Sciences 8 (15): 104-25.

Granda, María J., and Patricio Yánez M. 2017. *"Study on the perception of the benefits of the forest-leisure program in the Ecuadorian Amazon region."* La Granja: Revista de Ciencias de la Vida Vol 26 (2): 28-37.

Griscom, Bronson W, Justin Adams, Peter W. Ellis, Richard A. Houghton, Guy Lomax, Daniela A. Miteva, William H. Schlesinger et al. 2017. "Natural climate solutions." *Proceedings of the National Academy of Sciences* 114 (44): 11645-650.

Guedez, Pierre Yves, and Bruno Guay. 2018. *"Ecuador's Pioneering Leadership on REDD+; A Look Back at UN-REDD Support Over the Last 10 Years."* Accessed September 17. https://www.un-redd.org/post/2018/09/04/ecuadors- pioneering-leadership-on-redda- look-back-at-un-redd-support-over-the-last-10- years.

Guzmán, Sandra, Mariana Castillo, and Alin Moncada. 2017. *"Financing efforts against climate change in latin america."* Politics, Globality, and Citizenship (6): 65-98.

Guzmán, Sandra. 2021. "*Climate finance is a key part of the COP26 negotiations*". International finance is of vital importance to ensure compliance with climate goals. Accessed December 6, 2021. https://redaccion.lamula.pe/2021/11/09/sandra-guzman-el-climate-finance-is-key-piece-of-the-negotiations-at-COP26/albertoniquen/.

Hayes, Tanya, Felipe Murtinho, and Wolff Hendrik. 2017. "The impact of payments for environmental services on communal lands: An analysis of the factors driving household land-use behavior in Ecuador." *World Development* 93 (4): 427-46.

Herrán, Claudia. 2012. *"UN-REDD Program: the contribution of developing countries to curb climate change."* Friedrich Ebert Stiftung (162): 1-7.

Hirsch, Thomas. 2018. *"Towards ambitious implementation of the Paris Agreement."* ACT Alliance 150: 3-56.

. 2014. "REDD+ highlights tenure problems, but does not solve them." *Center for International Forestry Research, CIFOR*.

ICLEI. 2020. "*Climate Finance* Glossary.*"* Accessed February 6, 2021. https://americadosul.iclei.org/wp-.
content/uploads/sites/78/2021/04/glossario-tap-en-v4.pdf

INEC. 2006. "*La pobalción indígena del Ecuador. Análisis de*

Estadísticas Socio- Demográficas". Chisaguano: 26-68.

IEA. 2016. "*Operating agent: Building Research Establishment, Garston.*" International Energy Agency. Key CO.

IPCC. 2013. *"Glossary. Climate Change 2013. Physical Basis: Contribution of Working Group I to the Fifth Assessment Report of the Intergovernmental Panel on Climate Change."* Cambridge, UK and New York: Cambridge University Press.

. 2014. *"Climate Change 2014: Synthesis Report. Contribution of Working Groups I, II and III to the Fifth Assessment Report of the Intergovernmental Panel on Climate Change.*" Geneva.

Jones, Kelly W., Nicolle Etchart, Margaret Holland, Lisa Naughton-Treves, and Rodrigo Arriagada. 2020. "*The impact of paying for forest conservation on perceived tenure security in Ecuador*." Conservation Letters 13 (4): 68-89.

Kambire, Hermann, Ida Nadia Djenontin, Augustin Kaboré, Houria Djoudi, Michael Balinga, Mathurin Zida, Samuel Assembe-Mvondo, and Maria Brockhaus. 2016. *"REDD+ efficiency, its effectiveness and equity[1]."* In The Context of REDD+ and adaptation to climate change in Burkina Faso: Drivers, agents and institutions, by Hermann Kambire, Ida Nadia Djenontin, Augustin Kaboré and Houria Djoudi. Center for International Forestry Research 7.

Kill, Jutta. 2017. *"From REDD+ projects to jurisdictional REDD+: more bad news for climate and communities."* World Rainforest Movement (231): 77-98.

. 2014. *"The new REDD move: from forests to landscapes more of the same, but bigger and with higher risks*." World Rainforest Movement (204): 3- 16.

Larrea, C, S Latorre, and R Burbano. 2017. *"Análisis multicriterial sobre alternativas para el desarrollo en la Amazonia en ¿Está agotado el periodo petrolero en Ecuador?"*. Quito: Ediciones La tierra. Universidad Andina Simón Bolívar.

Leonard, Stephen, and Christopher Martius. 2021. *"A current analysis of REDD+ performance payments from the Green Climate Fund. Suggestions for a system that pays for emissions reductions that are real and permanent*." Center for International Forestry Research, CIFOR.

León Paz, Julio Ramiro, and Rivera Amanda. 2020. "*Tenure Illegality and Inequality in Land Distribution in Ecuador as Conditions of Vulnerability."* Geopauta, 4 (1): 34-48.

Lohmann, Larry. 2020. "*The Green Climate Fund (GCF) must say No to more REDD+ funding requests.*

more REDD+ funding requests". World Rainforest Movement.

Lorenzo, Cristian. 2016. "*The Global Environment Facility (GEF) as a political-environmental actor in Latin America."* IDICSO Institute for Social Science Research (6): 14-24.

Lovejoy, Thomas, and Carlos Nobre. 2018. *"Amazon tipping point."* Advances Science 4: 47-78.

Lovera-Bilderbeek, Simone. 2019. "REDD+ and the Green Climate Fund: worst fears confirmed." *Global Forest Coalition* (17):79-199.

Macchi, Mirjam, Gonzalo Oviedo, Sarah Gotheil, Katherine Cross, Aghi Boedhihartono, Caterina Wolfangel, and Matthew Howell.

2008. "Indigenous and Traditional Peoples and Climate Change." *International Union for the Conservation of Nature.*

Matta, Jagannadha Rao, and Laura Schweitzer Meins. 2012. "International journal of forestry and forest industries." *International journal of forestry and forest industries. Unasylva* 63, (239): 2-79.

Ministry of Environment. 2017a. "*Deforestación Del Ecuador Continental Periodo 2014-2016".* 04 July.

http://190.152.46.74/documents/10179/1149768/DEFORESTACION _ECUADOR_CONTINENTAL_21%204_2016.pdf/8f5a1064-4aa7- 47b0-80a0-3a54bbbb9fae . 2019. "Socio Bosque Project." Accessed May 17, 2021. https://www.ambiente.gob.ec/wp-content/uploads/downloads/2020/07/12.SOCIO_BOSQUE.pdf.

. 2019a. "Forests for Good Living - REDD+ Ecuador". Second Summary of Information on Addressing and Respecting Safeguards for REDD+ in Ecuador.

. 2019b. "First Nationally Determined Contribution to the Paris Agreement under the United Nations Framework Convention on Climate Change ". Quito-Ecuador: 4-50.

. 2019c. *"REDD+ Action Plan"* Accessed March 8. http://reddecuador.ambiente.gob.ec/redd/plan-de-accion-redd/.

. 2020. *"Deforestation and Regeneration at the Provincial Level of the Period 2016--2018 of Continental Ecuador. Interactive Environmental Map. Unique System of Environmental Indicators. SUIA."* Accessed May 6. http://ide.ambiente.gob.ec/mapainteractivo/.

Deforestation and Regeneration at the Provincial Level for the Period 2016-2018 of Continental Ecuador. Interactive Environmental Map. Unique System of Environmental Indicators. SUIA." Accessed May 6. http://ide.ambiente.gob.ec/mapainteractivo/.

. 2022. "*Deforestation and Regeneration at the Provincial Level of the Period 2016-2018 of Continental Ecuador. Interactive Environmental Map. Unique System of Environmental Indicators. SUIA*." Accessed April 19. http://ide.ambiente.gob.ec/mapainteractivo/.

. 2023. *"Ecuador promotes forest conservation through REDD+ Roundtable."* Accessed May 1. https://www.ambiente.gob.ec/ecuador-promueve-la-conservacion- de-los-bosques-a-traves-de-la-mesa-redd/

Molina, Mario, Julia Carabias, and José Sarukhán. 2017. *"El cambio climático: causas, efectos y soluciones"."* Mexico: Fondo de Cultura Económica.

Montaño, Doménica. 2021. *"Nuevo estudio: en los últimos 26 años Ecuador ha perdido más de 2 millones de hectáreas de bosque"."* Amazonia Socioambiental (18): 46- 72.

Montero, Maritza. 2002. *"Construcción del otro, liberación de sí mismo."* Utopia y praxis latinoamericana: revista internacional de filosofia iberoamericana y teoría social (16): 41-52.

. 2004. "*Empowerment in the community, its difficulties and* scope." Psychosocial Intervention 13.1: 5-19.

. 2010. *"Strengthening citizenship and social transformation: area of encounter between political psychology and community psychology."* Psykhe (Santiago) 19.2: 51-63.

Moreano, Melissa. 2012. *"Socio Bosque and Green Capitalism."* Línea de Fuego.

. 2014. "Cash for Conservation how does Socio Bosque work?". Terra Incognita (88): 31-5.

Muñoz, Magdalena. 2021. *"Mesa de Trabajo REDD+: 8 años en la preparación e implementación de REDD+ en Ecuador."* PROAmazonia. https://www.proamazonia.org/mesa-de-trabajo-redd- 8-anos-en-la-preparacion-e-implementacion- de-redd-en-ecuador/

Myers, Rodd, Anne M. Larson, Ashwin Ravikumar, Laura F. Kowler, Anastasia Yang, and Tim Trench. 2018. *"Messiness of forest governance: how technical approaches suppress politics in REDD+ and conservation projects."* Global Environmental Change (50): 314-24.

United Nations. 2015. "Paris Agreement" Accessed May 5, 2021. https://unfccc.int/files/meetings/paris_nov_2015/application/pdf/pari s_agreement_spanish_.pdf.

Naranjo, Alex. 2018. *"Ecuador: peoples, communities and nature in the face of oil palm."* Word Rainforest Movement (240).

Nemirovsky, Yanina. 2019. *"Tell me where the money goes and I'll tell you if you'll meet your climate goals."* Connectas (6): 3-9.

Nepstad, Daniel. Juan Pablo Ardila, María de los Ángeles Barrionuevo, Andrea Garzón, Juan Gabriel Rojas, Rafael Vargas, Jonah Busch, Eduardo Bedoya Garland, and Tathiana Bezerra. 2019. *"Evaluation of the Impact of public policies aimed at reducing deforestation and degradation and actions aimed at sustainable forest management in Ecuador."* (1): 15-233.

Norman, Marigold, and Smita Nakhooda. 2015. *"The state of REDD+ finance"."* Center for Global Development Working Paper 378.

OECD. 2015. *"Climate Finance in 2013-14 and the USD 100 billion Goal."* Organisation for Economic Cooperation and Development (OECD) in collaboration with Climate Policy Initiative (CPI).

OECD. 2019. *"Financing Climate Futures: Rethinking Infrastructure."* The World Bank United Nations Environment Programme 5.

Olca. 2015. *"Call to Action to reject REDD and extractive industries."* Latin American Observatory of Environmental Conflicts.

Olesen, Asger, Hannes Bottcher, Anne Siemons, Lara Herrmann, Christopher Martius, Rosa María Román Cuesta, Stibniati Atmadja et al. 2018. *"Study on EUfinancing of REDD+ related activities, and results-based payments pre and post 2020: Sources, costeffectiveness and fair allocation of incentives."* COWI.

Ordenavia, Noelia, Paola Bernal, and Winnie Narvaez. 2021. "Illustrations of the REDD+ mechanism. sustainable development or sustain the development model?". March Issue (71): 19-31.

Overbeek, Winnie. 2020. *"Indonesia: REDD+, European development finance and the 'low carbon economy'."* World Rainforest Movement (252): 39-106.

Paz Cardona, Antonio José. 2019. *"New report reveals that northern Ecuadorian Chocó has lost 61% of its forests.""* Mongabay (31): 6-25.

. 2022. "*The Ecuadorian Amazon has lost more than 623,000 hectares in two decades"."* Mongabay (17): 9-32

UNDP. 2012. *"Preparedness for Climate Finance. A Framework for Understanding What It Means to Be Ready to Use Climate Finance'"* UNDP. Vandeweerd, Veerle, Yannick Glemarec, and Simon Billett.

. 2019. *"Ecuador receives US$ 18.5 million for having reduced its deforestation""* Accessed July 09. https://www.climateandforests-undp.. org/node/5576

UNEP. 2018. *"REDD+ Academy Learning Diary. Forests and Climate Change."* Nairobi-Kenya 6 (3).

. 2018a. *"REDD+ Academy Learning Diary. Drivers of Deforestation and Degradation.""* Nairobi-Kenya 6 (3).

. 2018b. *"REDD+ Academy Learning Diary. The REDD+ Initiative and the UNFCCC."* Nairobi-Kenya 7 (3).

PROAmazonia. 2021. *"Mesa de Trabajo REDD+: 8 años en la preparación e implementación de REDD+ en Ecuador" (REDD+ Working Group: 8 years in the preparation and implementation of REDD+ in Ecuador)".* Accessed March 29. https://www.proamazonia.org/mesa-de-trabajo-redd-8- anos-en-la-preparacion-e- implementacion-de- redd-en-ecuador/.

Mesa de Trabajo REDD+: 8 años en la preparación e implementación de REDD+ en Ecuador" (REDD+ Working Group: 8 years in the preparation and implementation of REDD+ in Ecuador). Accessed March 29. https://www.proamazonia.org/mesa-de-trabajo-redd-8-anos-en-la- preparacion-e-implementacion-de-redd-en-ecuador/.

. 2021b. *"What is PROAmazonia?""* Accessed March 5.

https://www.proamazonia.org/en/inicio/que-es-proamazonia/.

. 2024. *"PROAmazonia Management Report""* Accessed April 17.https://info.undp. org/ docs/pdc/Documents/ECU/Annual%20Pro gress%20Project%20Report_2020.pdf.

Ramos, Tomás B, Inés Alves, Rui Subtil, and João Joanaz de Melo. 2007. "*Environmental performance policy indicators for the public sector: The case of the defence sector*". Journal of Environmental management 82, (4): 410-432.

Ramírez, Alonso. 2016. *"REDD+ and Costa Rican forest governance"."* Studies in Political Ecology, Development and Social Change (61): 219.

Ramos, Tomás B, Inés Alves, Rui Subtil, and João Joanaz de Melo. 2007. *"Environmental performance policy indicators for the public sector: The case of the defence sector."* Journal of Environmental management 82, (4): 410-432.

Romano, Antonio, Giuseppe Scandurra, Alfonso Carfora, and Monica Ronghi. 2018. "*Climate Finance as an Instrument to Promote the Green Growth in Developing Countries*." Rome-Italy: Springer.

Romijn, Erika, Celso B. Lantican, Martin Herold, Erik Lindquist, Robert Ochieng, Arief Wijaya, Daniel Murdiyarso, and Louis Verchot. 2015. *"Assessing change in national forest monitoring capacities of 99 tropical countries*." Forest Ecology and Management 352: 109-23.

Ramos, Melissa. 2022. "*A collective effort to solve Ecuador's land problem."* International Land Coalition, (7), 3-10.

Ron, Santiago. 2020. "Natural regions of Ecuador". *BIOWEB.* Edited by Pontificia Universidad Católica del Ecuador. Accessed 17 February 2023.
https://bioweb.bio/faunaweb/amphibiaweb/RegionesNaturales

Rudel, Thomas, Oliver Coomes, Emilio Moran, Frederic Achard, Arild Angelsen, Jianchu Xu and Eric Lambin. 2005. "*Forest transitions: towards a global understanding of land use change.*" Global environmental change 15 (1).

Salvador, Desiré. 2021. *"Latin America continues to rely on fossil fuels to generate income.""* Comunicar (71): 4-29.

Samaniego, Jose Luis, José Eduardo Alatorre, Orlando Reyes, Jimy Ferrer, Lina Muñoz, and Laura Arpaia. 2019. *"Overview of Nationally Determined Contributions in Latin America and the Caribbean, 2019: Progress towards compliance with the Paris Agreement""* (81): 28-51.

Sánchez, Euclides. 2000. *"Todos con la" esperanza": continuidad de la participación comunitaria"".* Comisión de Estudios de Postgrado, Facultad de Humanidades y Educación, Universidad Central de Venezuela.

Sean. 2013. *"Green Climate Fund Briefing Paper. Pre-Cop 19."* Regatta.

Sierra, Rodrigo, Oscar Calva, and Alejandra Guevara. 2021. *"Deforestation in Ecuador, 1900-2018. Promoting factors and recent trends.*" Ministry of Environment and Water of Ecuador, Ministry of Agriculture of Ecuador, in the framework of the implementation of the Amazonian Integral Program for Forest

Conservation and Sustainable Production (18): 2-216.

Skutsch, Margaret, and Esther Turnhout. 2018. "*How REDD+ Is Performing Communities.*" Forests 9, (10): 638.

Solís, Arturo. 2021. "*The 10 countries that pollute the planet the most".*" Forbes (5): 1-14.

Springate-Baginski, Oliver, and Eva Wollenberg. 2010. *"REDD, forest governance and rural livelihoods: the emerging agenda.*" CIFOR.

Stern, Nicholas, Siobhan Peters, Vicki Bakhshi, Alex Bowen, Catherine Cameron, Sebastian Catovsky and Diane Crane. 2006. "*Stern Review: The economics of climate change.*" Cambridge: Cambridge University Press.

Streck, Charlotte. 2020. "*Who Owns REDD+? Carbon Markets, Carbon Rights and Entitlements to REDD+ Finance.*" Forests 11 (9): 959.

Suarez, Victoria. 2017. *"Communities, indigenous peoples and decision makers, key actors in the Summary of Information from the REDD+ Working Table.*" UN-REDD Programme.

Suarez, Victoria. 2017. *"Communities, Indigenous Peoples and decision-makers, key actors in the REDD+ Working Table Information Summary."* UN-REDD Programme.

Sunderlin, William, Anne Larson, and Peter Cronkleton . 2010. *"Forest tenure rights and REDD+: From inertia to policy solutions."* In Angelsen, Brockhaus; Kanninen, Markku; Sunderlin, William; Wertz-Kanounnikoff, Sheila, from The Implementation of REDD+ National Strategy and Policy Options. Indonesia: CIFOR (5): 124-90.

Sunderlin, William, Andini Desita Ekaputri, Erin O. Sills, Amy E. Duchelle, Demetrius Kweka, Rachael Diprose, Nike Doggart, et al. 2014. *"Insightsfrom 23 Subnational Initiatives in Six Countries. from The Challenge of Establishing REDD+ on the Ground."* CIFOR (104).

Sunderlin, William, and Stibniati Atmadja. 2009. "*Is REDD+ and idea whose time has come, or gone?*". Realising REDD: national strategy and policy options. Bogor: CIFOR.

Sweeney, Gareth, Rebecca Dobson, Krina Despota, and Dieter Zinnbauer. 2011. "*Defining the Challenge: Threats to Climate Governance Effectiveness.*" In Global Corruption Report: Climate Change, by Alyson Warhurst Maplecroft. London: Earthscan (62): 11-29.

Takaki, Francisco Takaki. 2010. "*Basic Information for the Construction of the Deforestation Rate*". General Directorate of Geography. National Institute of Statistics and Geography. Mexico.

Ulloa, Astrid. 2013. *"Controlling nature: transnational environmentalism and local negotiations around climate change in indigenous territories in Colombia.*" Iberoamericana: 117-33.

Ulloa, Janette, and Rodrigo Sierra. 2019. *"Preliminary Assessment of Deforestation Management Gaps in Ecuador in the Early 2020s."* PROAmazonia - UNDP 25.

Vásquez, Edwin. 2013. *"Pueblos Amazónicos expusieron visión propia sobre REDD+ en Foro Permanente"*. SERVINDI. Intercultural communication for a more humane and diverse world.

Vatn, Arild, and Pál Vedeld. 2011. *"Getting Ready!: A Study of*

National Governance Structures for REDD+." Noragric Report 59.

Vegar, Bárd, Pablo Gutman, Charlie Parker, and Kristina Van Dexter. 2013. *"WWF Guide to Building REDD+ Strategies. Resources and tools for global REDD+ practitioners*." WWF Forests and Climate Program.

Vijge, Marjanneke J, Maria Brockhaus, Monica Di Gregorio, and Efrian Muharrom. 2016. *"Framing national REDD+ benefits, monitoring, governance and finance: A comparative analysis of seven countries."* Global Environmental Change (39): 57-68.

Warmikuna, Samanta. 2016. *"Working Paper on the Sapara Nation, its history and a genocide in the making."* Ecological Action.

Watson, Charlene, and Liane Schalatek. 2019. *"Climate finance thematic overview: Financing for REDD+. Background information on climate finance*." Climate Funds Update (5): 1-5.

. 2021. *"Regional overview of climate finance: Latin America".* Climate Funds Update. Heinrich Bull Stiftung- North America ODI.

Watson, Charlene, Liane Schalatek and Aurélien Evéquoz. 2022. *"Regional climate finance report: Latin America*." Climate Funds Update (6): 1-7.

Wong, Grace Yee, Cecilia Luttrell, Lasse Loft, Anastasia Yang, Thuy Thu Pham, Daisuke Naito, Samuel Assembe-Mvondo, and Maria Brockhaus. 2019. *"'Narratives in REDD+ benefit sharing: Examining evidence within and beyond the forest sector."'* Climate Policy 19, (8): 1038-1051.

Xu, Xibao, Guishan Yang, Yan Tan, Qianlai Zhuang, Xuguang Tang, Kaiyan Zhao, and Sirui Wang. 2017. *"Factors influencing industrial carbon emissions and strategies for carbon mitigation in the Yangtze River Delta of China."* Journal of cleaner production (142): 360716.

Zürn, Michael. 2018. *"A theory of global governance: authority, legitimacy, and contestation."* Oxford University Press.

Zúñiga, Nieves. 2022. *"Ecuador - Context and Land Governance." Instituto* de Altos Estudios Nacionales - Land (15): 7-13.

Annexes

Annex 1: List of interviewees

1. **Arild Angelsen** - Researcher and Author of REDD+ Works
Professor of Economics at the Norwegian University of Life Sciences (NMBU), Senior Associate at CIFOR, Global Coordinator of the Poverty and Environment Network (PEN), Researcher and author of multiple works. Since 2007 he has focused his research on REDD+: global architecture (in particular national reference levels), national strategies and policies, and local project design and evaluation.

2. **Carolina Rosero** - Conservation International
Environmental Policy Manager, Conservation International

3. **Cristina García Soto** - WWF
REDD+ Focal Point in the Undersecretariat of Climate Change of the Ministry of the Environment
Currently: Forest and Water Programme Officer at WWF

4. **David Romo Vallejo** - Universidad San Francisco de Quito
Director of the Ethnic Diversity Program
Member of the REDD+ Working Group in the second and third phases.

5. **David Yedra** - GAD Provincial de Pastaza
Director of Environmental Management of the Pastaza GAD

6. **Duval Llaguno Ribadeneira** - Inter-American Development Bank

Inter-American Development Bank Natural Resources Specialist

7. **Francisco Moscoso Silva** - PROAmazonía
Technical Specialist in Monitoring and Follow-up of REDD+ Measures and Actions Conducted the update of the Financial Strategy of the REDD+ AP.

8. **Jaime Toro** - Nature and Culture International
Pastaza Mosaic Coordinator

9. **Jessica Gallegos** - Ministry of Environment, Water and Ecological Transition
MAATE Climate Change Mitigation Specialist

10. **Manuel Shiguango** - CONFENIAE / UN REDD+
Territorial Technician

11. **Patricia Serrano Roca** - PROAmazonia
PROAmazonia Program Manager at UNDP

Annex 2: Working Table organizations

Sector 1	Civil Society
Academy	1. **Amazon State University (UEA)** 2. **University of San Francisco de Quito (USFQ)** 3. **Pontifical Catholic University of Ecuador (PUCE)** 4. **Private Technical University of Loja (UTPL)**
NGO **National**	5. **Ecuadorian Committee for the Defense of Nature and Environment** 6. **Conservation International - CI** 7. **National Working Group on Voluntary Forest Certification in Ecuador (CEFOVE)** 8. **Heifer Foundation** 9. **World Wild Fund - WWF** 10. **Altrópico Foundation** 11. **Ceiba Foundation** 12. **International Network for Bamboo and Rattan - INBAR** 13. **Wildlife Conservation Society** 14. **Nature and Culture International** 15. **Pachamama Foundation**
Sector 2	**Civil Society**

Women's and youth organizations	16. **Association of Waorani Women of the Ecuadorian Amazon (AMWAE)** 17. **CONFENIAE (Commission on Women and Health, Family and Nutrition)** 18. **La Chakra Producers Association** 19. **Network of Young Environmentalists of Southern Ecuador (Red JASE)**
Women's and youth organizations	**20 Confederation of Indigenous Nationalities of the Ecuadorian Amazon (CONFENIAE)** 5. **Quijos Orginary Nation (NAOQUI)** 6. **Provincial Federation of the Shuar Nationality of Zamora Chinchipe (FEPNASH-SCH)** 7. **Association of Shuar Centers of Santiago** 8. **Association of Kichwa Women of Napo (AMUKINA)** 9. **Shiwuar Foundation Without Borders (FCAE)**
Indigenous organizations in the Sierra	16. **Interprovincial Federation of Indigenous Saraguros (FIIS)** 17. **Federation of Awá Centers of Ecuador (FCAE)**
Montubio and peasant organizations	20. **Federation of Montubio Organizations of Ecuador (FEDOMEC)** 21. **Northwestern Union of Peasant Organizations and Populations of Pichincha (UNOCYPP)**

Communities premises	**30. Network of Social and Community Organizations in Water Management in Ecuador - ROSCGAE** **31. Network of Socio Bosque Communities of Napo** **32. Association of Forests and Moorlands for Life Imbabura** **33. Shuar Yumisim Community - PSB**
Sector 3	**Private Sector**
Guilds **National**	**34. Ecuadorian Association of Wood Industrialists -** **AIMA**
Associations of small	**35. Cacao y Chocolate de Napo Consortium** **36. Charolais Association of Morona Santiago**

producers	10. **Agro Artisanal Association of Ecological Producers - APECAP** 11. **Federation of Small Organic Agricultural Exporters of the Southern Ecuadorian Amazon - APEOSAE** 12. **Producers Association - ASOSUMACO** 13. **Union of Agricultural Producers of Morona Santiago**
Company **Private**	**41. Canandé Green**
Sector 4	**Guest groups**
Projects or programs implementing REDD+ actions	18. **Mancomunidad Bosque Seco** 19. **Consorcio Público Bosque Petrificado Puyango Public Consortium** 20. **Consorcio GS Agroforestry Consortium San Pablo del Lago** 21. **Loja Fair Corporation**

Printed by Books on Demand GmbH, Norderstedt / Germany